Musculoskeletal Health in Women

Elinor Mody • Elizabeth Matzkin
Editors

Musculoskeletal Health in Women

Springer

Editors
Elinor Mody, MD
Department of Rheumatology
Gretchen and Edward
Fish Center for Women's Health
Brigham and Women's Hospital
Chestnut Hill, MA
USA

Elizabeth Matzkin, MS, MD
Department of Orthopaedic Surgery
Brigham and Women's Hospital
Boston, MA
USA

ISBN 978-1-4471-4711-4 ISBN 978-1-4471-4712-1 (eBook)
DOI 10.1007/978-1-4471-4712-1
Springer London Heidelberg New York Dordrecht

Library of Congress Control Number: 2013953724

Printed on acid-free paper

Springer is part of Springer Science+Business Media (www.springer.com)

This book is dedicated to David and Hannah Lerner, Vinod Mody, MD, and Nell Mody. You are the best family anyone could ever have.

Elinor Mody, MD

To all the special women in my life – you know who you are! To my mother, sisters, friends, and my three beautiful, intelligent, and athletic daughters – Abigail, Samantha, and Emily. Lastly, to my husband, who supports me in all that I do.

Elizabeth Matzkin, MS, MD

Preface

It has been 40 years since Title IX was passed, giving females equal opportunities for sports participation at federally funded programs. The number of women participating in sports has skyrocketed ever since.

As more women participate in sports and lead active lifestyles, we need to understand that they cannot train and be treated the same as men. Females are at higher risk for several medical problems compared to their male counterparts. We need to identify these risks, treat them, and, more importantly, learn how to prevent them.

Taking care of the female athlete or women's musculoskeletal problems requires a multidisciplinary understanding of the issues. Women's musculoskeletal health is an emerging body of knowledge, and it is important for health-care providers to understand this and be well versed in this area.

Musculoskeletal diseases are the leading cause of disability in the United States. Health-care costs are rising at an alarming trajectory. We need to educate ourselves to prevent potential musculoskeletal problems before they arise. If we can educate young active women to optimize their bone health, imagine how many insufficiency fractures we can prevent when they are older!

The collaboration of numerous health-care providers from different specialties presenting and covering different topics pertaining to women's musculoskeletal health in this book is just a beginning. We hope this book opens many doors to pursue better diagnostic, treatment, and prevention strategies for us all in the future.

Chestnut Hill, MA, USA
Boston, MA, USA

Elinor Mody, MD
Elizabeth Matzkin, MS, MD

Contents

Contributors

Jennifer Baima, MD Division of Women's Health, Department of Medicine, Harvard Medical School, Brigham and Women's Hospital, Chestnut Hill, MA, USA

Nicholas Colacchio, MD Department of Orthopaedics, Tufts Medical Center, Boston, MA, USA

Emily J. Curry, BA Department of Orthopaedic Surgery, Brigham and Women's Hospital, Boston, MA, USA

Dean Ehrlich, BS Department of Orthopaedics, Tufts University School of Medicine, Tufts Medical Center, Boston, MA, USA

Jingyi Gong, BS Department of Molecular Cellular Biology/Biochemistry, Brown University, Providence, RI, USA

Elizabeth Matzkin, MS, MD Department of Orthopaedic Surgery, Brigham and Women's Hospital, Boston, MA, USA

Elinor Mody, MD Department of Rheumatology, Gretchen and Edward Fish Center for Women's Health, Brigham and Women's Hospital, Chestnut Hill, MA, USA

Nicholas K. Muraoka, DO Department of Physical Medicinal and Rehabilitation, Spaulding Rehabilitation Hospital, Boston, MA, USA

Kentaro Onishi, DO Department of Physical Medicine and Rehabilitation, University of California, Irvine Medical Center, Orange, CA, USA

Gabrielle M. Paci, BA Department of Orthopaedic Surgery, Center for Advanced Orthopaedic Studies, Harvard Medical School, Boston, MA, USA

Margaret Seton, MD Department of Rheumatology, Immunology, and Allergy, Brigham and Women's Hospital, Boston, MA, USA

Eric L. Smith, MD Department of Orthopaedics, Tufts Medical Center, Boston, MA, USA

Alexis M. Ziemba, BA Department of Anesthesia, Pain, and Critical Care, Massachusetts General Hospital, Boston, MA, USA

Chapter 1
The Female Athlete Triad

Elizabeth Matzkin and Gabrielle M. Paci

Abstract The female athlete triad describes a spectrum of eating behaviors that result in low energy availability, accompanied by a range of less than ideal menstrual function and bone mineral density. Though the true prevalence of the triad along its full spectrum of interrelated pathologies remains unclear, limited studies have reported prevalence in 0–40 % of athletes and 0–3 % of controls. Diagnosis of the female athlete triad relies on a high index of suspicion among primary care physicians and sports specialists alike. If there is concern for one component of the triad, workup for the other components should be performed. Prevention is critical in high-risk groups. The mainstay of treatment is to increase energy availability. A multidisciplinary approach to treatment is best, with involvement of physicians from primary care, psychiatry, and sports specialists as needed, in addition to a nutritionist, trainer, and family members. Future investigation should focus on better quantifying the problem and options for intervention.

Keywords Female athlete triad • Athlete health • Energy availability • Disordered eating • Menstrual dysfunction • Amenorrhea • Bone mineral density • Osteoporosis

Abbreviations

AN	Anorexia nervosa
BMD	Bone mineral density
BN	Bulimia nervosa

E. Matzkin, MS, MD (✉)
Department of Orthopaedic Surgery, Brigham and Women's Hospital,
75 Francis Street, Boston, MA 02115, USA
e-mail: ematzkin@partners.org

G.M. Paci, BA
Department of Orthopaedic Surgery, Center for Advanced Orthopaedic Studies,
Harvard Medical School, 99 Brookline Ave, RN 115, Boston, MA 02215, USA
e-mail: gabrielle_paci@brown.edu

E. Mody, E. Matzkin (eds.), *Musculoskeletal Health in Women*,
DOI 10.1007/978-1-4471-4712-1_1, © Springer-Verlag London 2014

FAT	Female athlete triad
GnRH	Gonadotropin-releasing hormone
LH	Luteinizing hormone
OCP	Oral contraceptive pill

Introduction

Since the advent of Title IX, passed by Congress in 1972, female representation among high school athletes has increased from 7 to 41 % in 2009 [1]. This education amendment, which rejected the ability for federally funded schools to exclude athletes on the basis of sex, has clearly resulted in benefits for young female athletes. Among such benefits are increased self-esteem, higher high school graduation rates, and overall improved health [1]. Since this time, however, the prevalence of eating disorders among female athletes has also become apparent [2]. Most commonly seen in those who train in sports that thrive on low body mass, including gymnastics, ballet, swimming, diving, and running, the prevalence of eating disorders in this high-risk group is reported as high as 42 % [2, 3]. However, eating disorders among female athletes were not identified as solitary findings. In 1992, the Task Force on Women's Issues of the American College of Sports Medicine (ACSM) called a consensus conference in Seattle, Washington, to discuss an observed association between disordered eating, amenorrhea, and osteoporosis in adolescent and young adult female athletes [4, 5]. They coined the term Female Athlete Triad (FAT) to describe these interrelated pathologies and published a position stand in 1997, aimed at guiding treatment as well as increasing prevention and risk reduction [5, 6].

The ACSM updated their position stand in 2007, further describing the relationship as it exists along a spectrum of eating behaviors that result in low energy availability, accompanied by a range of menstrual function and bone mineral density (BMD) that all fall under the umbrella of the diagnosis [6, 7] (Fig. 1.1). The ACSM's updated position stand emphasized the importance of the presence of any one component of the triad and called for further investigation into epidemiologic prevalence and outcome studies as well as ways to prevent, identify, and treat this condition. To this end, research has increased awareness of the condition among athletes and physicians alike. Cross-sectional studies performed in variety of populations have demonstrated prevalence of FAT ranging from 0 to 40 % of athletes and 0 to 3 % of controls [8–18] (Table 1.1). The prevalence of one or more components of the triad is much higher, reported as high as 78 % in the literature [8, 11, 13]. Much remains to be learned, however, with regard to actual prevalence among high-risk groups as well as optimal treatment and management of this multisystem triad.

Energy Availability

Energy availability, or the amount of energy available for the body to function following athletic training, is calculated as the difference between dietary energy intake and exercise-related energy expenditure [7]. In a state of low energy

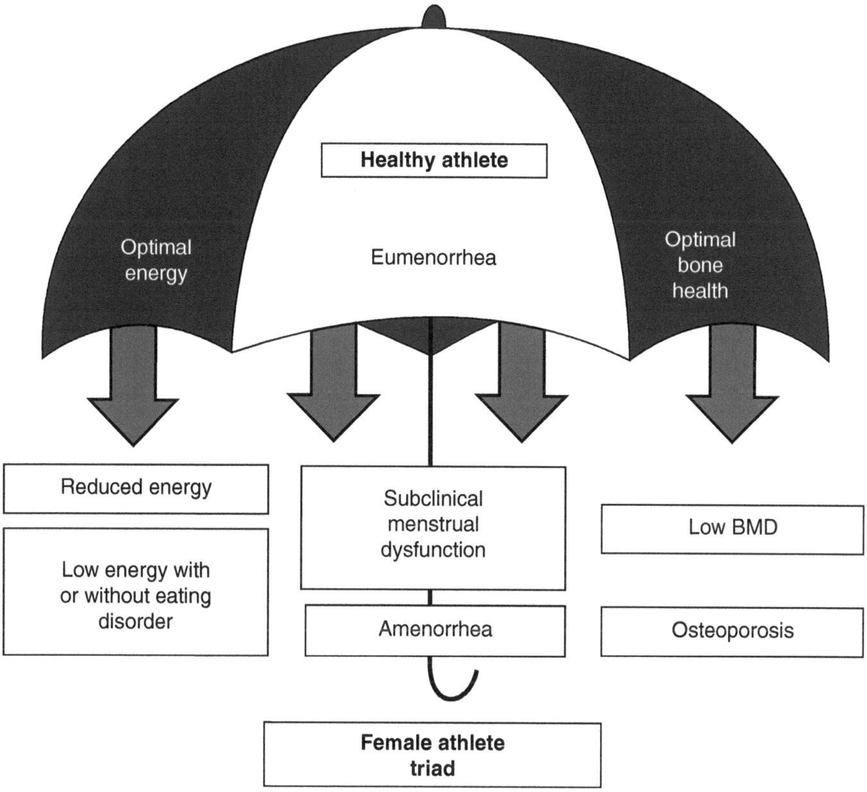

Fig. 1.1 A spectrum of energy availability, menstrual dysfunction, and bone mineral density fall under the umbrella diagnosis of the female athlete triad

availability, resources are redistributed to prioritize survival mechanisms including thermoregulation and cell maintenance at the expense of noncritical physiologic functions such as growth, fat storage, and menstruation [19]. Low energy availability can result from decreased dietary intake with or without an eating disorder, increased exercise expenditure, or a combination thereof.

It is important to distinguish disordered eating resulting in low energy availability from a diagnosis of an eating disorder such as anorexia nervosa (AN) or bulimia nervosa (BM). Disordered eating practices include a spectrum of subclinical behaviors that negatively impact health, ranging from restriction and purging to use of diet pills and laxatives [3, 20]. AN is a classified psychiatric condition characterized by restrictive or purging eating behaviors accompanied by body weight that is at least 15 % below ideal body weight and amenorrhea. BM, likewise, is a classified psychiatric condition consisting of binge eating followed by compulsive behaviors such as purging or excessive exercise, resulting in normal or slightly increased body weight. While disordered eating and low energy availability are common among female athletes, diagnosed eating disorders are low in prevalence [4]. An actual eating disorder is not required to achieve low energy availability and FAT may be present anywhere along the spectrum of energy imbalance. Female athletes may, as such, be entirely unaware that they are failing to meet their energy intake requirements.

Table 1.1 Prevalence of the female athlete triad

Study	Subjects	Disordered eating	Menstrual dysfunction	Low bone mineral density	Female athlete triad
Beals et al. (2006) [8]	112 Collegiate athletes	25 %[a]	26 %[b]	(10 %)[c]	2.7 % ($n=3$)
Coelho et al. (2013) [9]	24 adolescent tennis Players vs. 21 sedentary controls (Brazil)	87.5 % athletes vs. 9.5 % controls[d]	33.3 % athletes vs. 9.5 % controls[b]	25 % athletes vs. 33.3 % controls[c]	4.2 % athletes vs. 0 controls
Doyle-Lucas et al. (2010) [10]	15 professional adult dancers and 15 controls	N/A[a,d]	40 % dancers vs. 7 % controls[b]	N/A[e]	40 % dancers
Hoch et al. (2009) [11]	80 varsity high school athletes and 80 sedentary controls	36 % athletes vs. 39 % controls[d]	54 % athletes vs. 21 % controls[b]	16 % athletes vs. 30 % controls[c]	1 % athletes vs. 1 % controls
Hoch et al. (2007) [12]	15 triathlon team athletes	60 %[f]	40 %[b]	N/A[c]	0
Nichols et al. (2006) [13]	170 high school interscholastic athletes	18.2 %[a]	23.5 %[b]	21.8 %[c]	1.2 %
Pollock et al. (2010) [14]	44 elite endurance runners (United Kingdom)	N/A[a]	63 %[b]	41.5 %[c]	15.9 %
Quah et al. (2009) [15]	67 elite athletes (Malaysia)	89.2 %[a]	61.9 %[b]	21.6 %[c]	1.9 %
Schtscherbyna et al. (2009) [16]	78 elite swimmers (Brazil)	44.9 %[a]	19.2 %[b]	15.4 %[g]	1.3 %
Torstveit et al. (2005) [17]	186 elite athletes and 145 controls (Norway)	N/A[a]	N/A[b]	N/A[c]	4.3 % athletes vs. 3.4 % controls
Vardar et al. (2005) [18]	224 female athletes (Turkey)	16.8 %[a]	19.2 %[b]	N/A[c]	1.36 %

[a]Disordered eating determined by questionnaire
[b]Menstrual dysfunction determined by questionnaire
[c]BMD determined by dual-energy X-ray absorptiometry scores; low BMD defined as z-score < -1
[d]Disordered eating defined as low energy availability (<45 kcal/kg lean body mass/day) determined by food diary and exercise tally
[e]BMD determined by dual-energy X-ray absorptiometry scores; low BMD definition unspecified
[f]Disordered eating defined as low energy availability (calorie intake not meeting calorie expenditure) determined by food diary and exercise tally
[g]BMD determined by dual-energy X-ray absorptiometry scores; low BMD defined as z-score < -2

Due to the logistical difficulties of measuring exact energy intake and expenditure in an individual, few studies have reported on the prevalence of low energy availability among female athletes. In two studies examining adolescent athletes and controls, low energy availability (≤45 kcal/kg of fat free mass per day) was calculated in 36–87.5 % of athletes and 9.5–39 % of controls [9, 11]. More often, studies have focused on questionnaire screening tools to measure rates of clinical eating disorders and disordered eating. In a Norwegian study of 572 female elite athletes, 20 % met the criteria for an eating disorder and another 21 % were considered to be "at risk" for an eating disorder, compared to 14 and 9 % of female controls, respectively [2]. The numbers were even more staggering when broken down by sport categories, with 30 % of female weight class athletes and 42 % of female aesthetic athletes qualifying for eating disorders [2]. Another study of 425 female collegiate athletes reported AN in 3.3 %, BN in 2.4 %, and questionnaire scores qualifying them as "at risk" in 15–32 %, depending on the survey used [8]. Whether female athletes are actually at increased risk of disordered eating remains unclear, however, as a recent meta-analysis reviewing 22 studies in the literature found no difference in prevalence of disordered eating between athletes and controls [21].

Menstrual Function

The menstrual cycle describes a series of endocrine-regulated physiologic changes necessary for reproduction in female mammals. Regular menstrual function, with cyclic ovulation resulting in predictable menses near the median of every 28±7 days, relies on activity of the hypothalamic-pituitary-ovarian axis. At the level of the brain, the hypothalamus is responsible for pulsatile gonadotropin-releasing hormone (GnRH) release, which then signals the pituitary to release the gonadotropic hormones: luteinizing hormone (LH) and follicle-stimulating hormone. LH and follicle-stimulating hormone target the reproductive system at the ovaries and uterus, stimulating ovulation and a series of hormonal fluctuations involving estrogen and progesterone that result in either menstruation or pregnancy. Disruption at any level of this complex cascade can lead to menstrual dysfunction, or irregularity of the menstrual cycle, with or without infertility.

Historically, it was thought that a minimum body fat percentage was required in order to maintain the aforementioned cascade resulting in regular menstruation [22]. However, mounting evidence that there is no significant difference in body fat composition between eumenorrheic and amenorrheic athletes has disputed this theory [23–25]. Theories pointing at stress as the culprit have been similarly disproven [26]. Currently in favor is the metabolic fuels hypothesis that connects energy availability to menstruation by way of neuronal GnRH secretion [19, 24, 26, 27]. Low energy availability has been linked to non-pulsatile GnRH secretion resulting in disturbed LH release, explaining the low levels of LH seen in amenorrheic athletes [28]. A threshold of availability under which menstrual dysfunction will occur has been identified as 30 kcal/kg of lean body mass per day [24]. A role for signaling factors such as leptin and adiponectin has also been implicated but remains unclear [29, 30].

Table 1.2 Definitions of the spectrum of menstrual dysfunction

Oligomenorrhea	Menstrual intervals > 35 days
Primary amenorrhea	Never having had menses by age 15 in the presence of secondary sex characteristics or within 5 years of breast development occurring before age 10
Secondary amenorrhea	Loss of menses for > 90 days
Subclinical luteal phase suppression	Luteal phase < 11 days and/or low progesterone levels in the presence of menses
Anovulation	Absence of ovulation

The menstrual dysfunction related to low energy availability in FAT includes a noncontinuous spectrum ranging from oligomenorrhea to anovulation. Primary amenorrhea, secondary amenorrhea, and subclinical luteal phase suppression in the presence of menses, as defined in Table 1.2, all fall within this spectrum.

Prevalence of menstrual dysfunction in female athletes remains difficult to accurately quantify. This is, in part, due to frequent use of questionnaires for assessment [31]. It is further complicated by the fact that estimates often do not account for subclinical menstrual dysfunction, such as luteal phase suppression, that leads to eumenorrheic infertility [25, 31]. Studies have suggested that, even in recreational runners, total anovulation rates (with or without changes in menses) are as high as 78 % [32]. A number of studies report that elite leanness athletes have significantly higher rates of menstrual dysfunction when compared with non-leanness elite athletes, suggesting that prevalence varies by sport [15, 17, 33]. It has been estimated that up to 79 % of women participating in weight-bearing sports, such as ballet and running, have some menstrual dysfunction [27].

Bone Strength

Bone strength is a function of many factors, including bone mineral content, BMD, and bone quality determinants such as microarchitecture. Bone remodeling, in the form of bone formation and resorption, is a continuous process with the maximum increase occurring during adolescence and overall peak density occurring by the third decade of life [34]. In a healthy athlete, exercise-induced mechanical stress increases bone density, such that weight-bearing exercise in athletes (volleyball, gymnastics) results in a 5–19 % increase in BMD [35]. Changes in BMD seen along the spectrum of FAT likely result from a combination of nutritional deficits as well as hypoestrogenism.

Bone mineralization relies on calcium, phosphorous, and vitamin D supply. Recent research has shown that vitamin D concentrations were inadequate (<50 nmol/L) in 62 % of athletes in a Northern latitude population [36]. A study looking at only female gymnasts in Australia, ages 10–17, found that 72 % had calcium intake below that recommended for age and 83 % were vitamin D insufficient (<75 nmol/L) [37]. In female athletes with FAT, energy imbalance may make these deficiencies even more common.

Late menarche and menstrual dysfunction are associated with decreased BMD in both athletes and controls [17]. Hypoestrogenism, associated with menstrual dysfunction in athletes with low energy availability, is a well-established cause of bone demineralization and low BMD [7]. Estrogen acts at two levels to increase bone mineralization: binding at alpha receptors leads to increased osteoblast and decreased osteoclast function while, during puberty, stimulation of growth hormone increases insulin-like growth factor-I [38]. The hallmark of FAT that leads to decreased bone strength is a state of low estrogen, also implicated in the low BMD commonly seen in postmenopausal women [7, 31]. The role of progesterone has also been documented in women with luteal phase suppression, who experience BMD loss of 2–4 % per year [39, 40].

Some studies in the literature claim to have identified a direct association between low energy availability and decreased BMD that is independent of hypoestrogenism. Cobb et al. did not measure estrogen or progesterone levels in subjects and, therefore, may have failed to identify eumenorrheic subjects with hypoestrogenism [41]. De Souza et al. did measure estrogen levels in all subjects but failed to measure progesterone and categorized women as "estrogen replete" if they produced a "sufficient increase to result in menstrual bleeding" [32]. This methodology may have, similarly, failed to identify subjects who had estrogen fluctuations sufficient to achieve menses but estrogen and progesterone still low enough to effect bone health. Potential direct effects of low energy availability on bone remain unclear.

It is clear from the literature that BMD is lower in athletes with known menstrual dysfunction. One study identified low BMD in 38 % of amenorrheic athletes ($n=21$) compared to 11 % of eumenorrheic athletes ($n=18$), which was equal to the controls (11 %, $n=18$) [42]. The low BMD seen in female athletes with menstrual dysfunction puts them at increased risk of fracture that is two to four times that seen in the eumenorrheic cohort [31].

Diagnosis

Given the complex, interrelated nature of FAT and frequent lack of apparent signs or symptoms, diagnosis can be a challenge and relies on a thorough history for identification of individual components. A high index of suspicion should be maintained for all female athletes. Prime opportunities for relevant screening include sports clearance physicals and annual exams [7]. Presentation of any one component of the triad or diagnosis of a commonly related pathology (stress fracture) should lead to full workup for FAT [7]. Initial signs and symptoms, if present, may be as nonspecific as fatigue, anemia, electrolyte abnormalities, and depression [3].

Clinicians should obtain a thorough history that includes diet and energy intake, eating behaviors, change in weight, exercise regimen, menstrual history, and history of sports-related injury including fracture [3]. In addition to height, weight, and vital signs, physical exam should include particular attention not to overlook signs of disordered eating, such as lanugo, Russell's sign on the dorsal hands, periodontal

Table 1.3 Laboratory tests for female athlete triad component workup

Energy availability
Electrolytes
Chemistry profile
CBC with differential
ESR
Thyroid function tests
Urinalysis
Menstrual function
Pregnancy test
LH
FSH
LH–FSH ratio
Prolactin
TSH
Bone mineral density
DXA scan

disease, and salivary gland enlargement [4]. Eating disorder surveys may help clinicians to identify patients at risk for disordered eating. The Eating Disorder Inventory, Eating Attitudes Test, and Eating Disorder Examination are valid instruments for such assessment [4]. Laboratory tests should be obtained to assess the patient's nutritional status as outlined in Table 1.3. Results within normal limits cannot rule out the possibility of disordered eating and, in the presence of concern, referral to a mental health practitioner should be made for further workup and diagnosis [7].

Menstrual dysfunction rarely presents with any symptoms beyond irregular menses and may be entirely asymptomatic, as in the case of eumenorrheic luteal phase suppression. When secondary amenorrhea is present, investigation into the cause is warranted, as FAT is considered root cause only through exclusion of the remaining differential diagnoses. Differential for secondary amenorrhea includes pregnancy, medications (oral contraceptives, intrauterine devices, antipsychotics), hypothalamic dysfunction (disordered eating, Kallmann's syndrome), pituitary dysfunction (prolactinoma, Sheehan's syndrome), ovarian dysfunction (menopause, premature ovarian failure, polycystic ovary syndrome), uterine dysfunction (Asherman's syndrome), and endocrine dysfunction (hypothyroidism, Cushing's disease). When primary amenorrhea is present, anatomic considerations including absent uterus must also be considered [3]. Laboratory tests should be performed for evaluation of menstrual dysfunction in all athletes, as outlined in Table 1.3 [7]. Other tests, including serum free testosterone, serum cortisol, and/or progesterone challenge (medroxyprogesterone acetate 10 mg daily for 7–10 days) may be performed according to signs and symptoms. If 3–6 months of treatment do not succeed in achieving menses, consultation with a reproductive specialist is recommended [7].

Athletes who experience any of a number of risk factors, including known menstrual dysfunction or disordered eating anywhere along the spectrum for 6 months or longer, should be assessed for low BMD [7]. Any history of stress fracture or fracture resulting from minimal trauma should also raise concern. Bone mineral density should be quantified using dual-energy X-ray absorptiometry and checked serially every 12 months in those with FAT [7]. Bone mineral density is a measure of bone remodeling balance that provides information about fracture risk. *Z*-scores represent standard deviation from the mean for a given age. The International Society for Clinical Densitometry recommends that, in premenopausal women, z-score < -2 should be considered "below the expected range for age" [43]. In athletes who participate in weight-bearing sports, where one would expect an above average BMD, a z-score between −1 and −2 should be of concern. The ACSM has further classified z-scores in the context of secondary clinical risk factors: in the presence of nutritional deficiency, hypoestrogenism, history of stress fracture, or other fracture risk, z-score < -1 is considered low BMD, while ≤ -2 is deemed osteoporosis [7].

Treatment

Ideally, FAT, would be addressed in a timely fashion that allows for prevention at its cornerstone. Multidisciplinary intervention, including oversight by a primary care physician, consultation with a nutritionist or dietician, psychiatric evaluation, and involvement of the patient's trainer and family, may aid in achieving improvement. The mainstay of treatment for FAT is increased energy availability with resumption of the menstrual cycle. In athletes who are under ideal weight, a weight goal should be established and achieved in weekly increments of 0.5–1 lb/week [3]. Energy intake should be monitored with a minimum goal of greater than 45 kcal/kg of fat free mass per day to achieve improvement in bone health [7]. Exercise should not be halted, but rather decreased by 10–20 % in the setting of weight and energy intake monitoring [3]. Nutrient intake should be assessed and supplemented where necessary, with a goal of 1,000–1,300 mg of calcium per day, 400–800 international units of vitamin D per day and 60–90 μg of vitamin K per day [7].

Though commonly used to treat osteoporosis in postmenopausal women, the role of estrogen supplementation in improving BMD for athletes with menstrual dysfunction remains controversial. Estrogen supplements, including hormone replacement therapy and/or oral contraceptive pills (OCPs), have been used to treat low BMD and osteoporosis secondary to athlete-associated amenorrhea despite the lack of supporting evidence in the literature [44]. Studies examining the effect of OCPs in athletes with functional hypothalamic amenorrhea over 8 months to >4 years have demonstrated equivocal increases in BMD compared to controls [45]. Cobb et al. performed the largest randomized controlled trial to date following patients over 2 years. Results were inconclusive due to a high rate of noncompliance among those randomized to OCPs as well as frequent spontaneous resumption of menses among the controls [46].

The pathophysiology of estrogen's effects on bone strength may help to explain its unclear role in women with menstrual dysfunction. Women with menstrual dysfunction experience decreased osteoblastic bone formation as well as increased osteoclastic bone resorption. Estrogen works primarily by inhibiting osteoclastic bone resorption, thereby only targeting one of the factors leading to low BMD in this population [44]. Furthermore, exogenous estrogens, such as OCPs, may inhibit the body's natural androgen secretion, which could contribute to an overall decrease in BMD [47].

Though estrogen therapies may halt bone loss by decreasing bone turnover, evidence suggests that they are unable to achieve increases to ideal BMD in this population [48]. It remains unclear whether preventing further decrease in BMD has any clinical benefit in preventing stress fractures [45]. Use of OCPs to achieve resumption of menses in FAT may lead to a false sense of security with regards to prognosis. In young athletes, the use of OCPs is further complicated by the potential for short stature, as the exogenous estrogen leads to premature closure of the growth plates [44]. As a result, current recommendations for use of OCPs suggest that this treatment only be used in patients over 16 years of age who have persistent FAT-related amenorrhea despite having achieved ideal nutrition and weight [7].

Bisphosphonates also remain controversial for use in the female athlete. Very little is known about their efficacy in treating premenopausal osteoporosis secondary to menstrual dysfunction and energy imbalance. Furthermore, their use is highly discouraged in premenopausal women due to the potential for teratogenic effects in the event of pregnancy [44].

Prognosis

Expected prognosis of FAT has not been well established. The decrease in BMD associated with FAT may not be fully reversible, even in the setting of resumption of menses [49–51]. Keen et al. found that, even after several years of having resumed menstrual cycles, former amenorrheic or oligomenorrheic athletes still exhibited only 85 % of the BMD measured in control athletes at mean 8-year follow-up [49]. A cohort of women with former intermittent menstrual dysfunction demonstrated an intermediate recovery of BMD (95 % of that measured in controls), implying that there may be a continuum of long-term consequences to bone that are related to severity [49].

Recent studies have investigated the possible association between FAT and cardiovascular health. Estrogen receptors within the vascular system are involved in mediation of endothelial-dependent vasodilation. In hypoestrogenic states, including the post-menopausal period and FAT, decreased levels of estrogen lead to endothelial dysfunction as an early sign of cardiovascular disease. This can be measured as decreased endothelium-dependent vasodilation of the brachial artery detected on ultrasound, which has been shown to be predictive of coronary endothelial dysfunction [52, 53]. Amenorrheic athletes have demonstrated lower endothelium-dependent vasodilation of the brachial artery when compared to both oligomenorrheic athletes and controls [54, 55]. One study identified endothelial dysfunction in 64 % ($n=14$) of professional dancers [56]. The association between coronary endothelial dysfunction and cardiovascular events has been well established, indicating that the menstrual dysfunction secondary to FAT does put patients at risk of events including myocardial infarction and stroke [57, 58].

Conclusion

The female athlete triad exists along a spectrum of decreased energy availability that results in a range of menstrual dysfunction and BMD. Individual components of the triad may be present in as many as 78 % of athletes, leading to long-term deleterious health effects. The mainstay of treatment is increasing energy availability through a combination of increased intake and modified exercise expenditure. Much remains to be learned, including prevalence of the triad components along the spectrum, as well as optimal strategies for management.

References

1. Thein-Nissenbaum JM, Carr KE. Female athlete triad syndrome in the high school athlete. Phys Ther Sport. 2011;12(3):108–16.
2. Sundgot-Borgen J, Torstveit MK. Prevalence of eating disorders in elite athletes is higher than in the general population. Clin J Sport Med. 2004;14(1):25–32.
3. Hobart JA, Smucker DR. The female athlete triad. Am Fam Physician. 2000;61(11):3357–64, 67.
4. Brunet 2nd M. Female athlete triad. Clin Sports Med. 2005;24(3):623–36, ix.
5. Yeager KK, Agostini R, Nattiv A, Drinkwater B. The female athlete triad: disordered eating, amenorrhea, osteoporosis. Med Sci Sports Exerc. 1993;25(7):775–7.
6. Otis CL, Drinkwater B, Johnson M, Loucks A, Wilmore J. American College of Sports Medicine position stand. The female athlete triad. Med Sci Sports Exerc. 1997;29(5):i–ix.
7. Nattiv A, Loucks AB, Manore MM, Sanborn CF, Sundgot-Borgen J, Warren MP. American College of Sports Medicine position stand. The female athlete triad. Med Sci Sports Exerc. 2007;39(10):1867–82.
8. Beals KA, Manore MM. Disorders of the female athlete triad among collegiate athletes. Int J Sport Nutr Exerc Metab. 2002;12(3):281–93.
9. Coelho GM, Farias ML, Mendonca LM, Mello DB, Lanzillotti HS, Ribeiro BG, et al. The prevalence of disordered eating and possible health consequences in adolescent female tennis players from Rio de Janeiro, Brazil. Appetite. 2013;64C:39–47.
10. Doyle-Lucas AF, Akers JD, Davy BM. Energetic efficiency, menstrual irregularity, and bone mineral density in elite professional female ballet dancers. J Dance Med Sci. 2010;14(4):146–54.
11. Hoch AZ, Pajewski NM, Moraski L, Carrera GF, Wilson CR, Hoffmann RG, et al. Prevalence of the female athlete triad in high school athletes and sedentary students. Clin J Sport Med. 2009;19(5):421–8.
12. Hoch AZ, Stavrakos JE, Schimke JE. Prevalence of female athlete triad characteristics in a club triathlon team. Arch Phys Med Rehabil. 2007;88(5):681–2.
13. Nichols JF, Rauh MJ, Lawson MJ, Ji M, Barkai HS. Prevalence of the female athlete triad syndrome among high school athletes. Arch Pediatr Adolesc Med. 2006;160(2):137–42.
14. Pollock N, Grogan C, Perry M, Pedlar C, Cooke K, Morrissey D, et al. Bone-mineral density and other features of the female athlete triad in elite endurance runners: a longitudinal and cross-sectional observational study. Int J Sport Nutr Exerc Metab. 2010;20(5):418–26.
15. Quah YV, Poh BK, Ng LO, Noor MI. The female athlete triad among elite Malaysian athletes: prevalence and associated factors. Asia Pac J Clin Nutr. 2009;18(2):200–8.
16. Schtscherbyna A, Soares EA, de Oliveira FP, Ribeiro BG. Female athlete triad in elite swimmers of the city of Rio de Janeiro, Brazil. Nutrition. 2009;25(6):634–9.
17. Torstveit MK, Sundgot-Borgen J. The female athlete triad exists in both elite athletes and controls. Med Sci Sports Exerc. 2005;37(9):1449–59.

18. Vardar SA, Vardar E, Altun GD, Kurt C, Ozturk L. Prevalence of the female athlete triad in Edirne, Turkey. J Sports Sci Med [Internet]. 2005;4:550–5. Cited 4 Mar 2013. Available from: http://www.jssm.org/vol4/n4/21/v4n4-21text.php
19. Wade GN, Schneider JE, Li HY. Control of fertility by metabolic cues. Am J Physiol. 1996; 270(1 Pt 1):E1–19.
20. Krentz EM, Warschburger P. A longitudinal investigation of sports-related risk factors for disordered eating in aesthetic sports. Scand J Med Sci Sports. 2013;23(3):303–10.
21. Coelho GM, Soares Ede A, Ribeiro BG. Are female athletes at increased risk for disordered eating and its complications? Appetite. 2010;55(3):379–87.
22. Frisch RE, McArthur JW. Menstrual cycles: fatness as a determinant of minimum weight for height necessary for their maintenance or onset. Science. 1974;185(4155):949–51.
23. Loucks AB, Horvath SM, Freedson PS. Menstrual status and validation of body fat prediction in athletes. Hum Biol. 1984;56(2):383–92.
24. Loucks AB. Energy availability, not body fatness, regulates reproductive function in women. Exerc Sport Sci Rev. 2003;31(3):144–8.
25. Redman LM, Loucks AB. Menstrual disorders in athletes. Sports Med. 2005;35(9):747–55.
26. Loucks AB, Verdun M, Heath EM. Low energy availability, not stress of exercise, alters LH pulsatility in exercising women. J Appl Physiol. 1998;84(1):37–46.
27. Warren MP, Perlroth NE. The effects of intense exercise on the female reproductive system. J Endocrinol. 2001;170(1):3–11.
28. Ackerman KE, Slusarz K, Guereca G, Pierce L, Slattery M, Mendes N, et al. Higher ghrelin and lower leptin secretion are associated with lower LH secretion in young amenorrheic athletes compared with eumenorrheic athletes and controls. Am J Physiol Endocrinol Metab. 2012; 302(7):E800–6.
29. Laughlin GA, Yen SS. Hypoleptinemia in women athletes: absence of a diurnal rhythm with amenorrhea. J Clin Endocrinol Metab. 1997;82(1):318–21.
30. Rodriguez-Pacheco F, Martinez-Fuentes AJ, Tovar S, Pinilla L, Tena-Sempere M, Dieguez C, et al. Regulation of pituitary cell function by adiponectin. Endocrinology. 2007;148(1):401–10.
31. Bennell K, Matheson G, Meeuwisse W, Brukner P. Risk factors for stress fractures. Sports Med. 1999;28(2):91–122.
32. De Souza MJ, Miller BE, Loucks AB, Luciano AA, Pescatello LS, Campbell CG, et al. High frequency of luteal phase deficiency and anovulation in recreational women runners: blunted elevation in follicle-stimulating hormone observed during luteal-follicular transition. J Clin Endocrinol Metab. 1998;83(12):4220–32.
33. Torstveit MK, Sundgot-Borgen J. Participation in leanness sports but not training volume is associated with menstrual dysfunction: a national survey of 1276 elite athletes and controls. Br J Sports Med. 2005;39(3):141–7.
34. Lambrinoudaki I, Papadimitriou D. Pathophysiology of bone loss in the female athlete. Ann N Y Acad Sci. 2010;1205:45–50.
35. Fehling PC, Alekel L, Clasey J, Rector A, Stillman RJ. A comparison of bone mineral densities among female athletes in impact loading and active loading sports. Bone. 1995;17(3):205–10.
36. Close GL, Russell J, Cobley JN, Owens DJ, Wilson G, Gregson W, et al. Assessment of vitamin D concentration in non-supplemented professional athletes and healthy adults during the winter months in the UK: implications for skeletal muscle function. J Sports Sci. 2013;31(4):344–53.
37. Lovell G. Vitamin D, status of females in an elite gymnastics program. Clin J Sport Med. 2008;18(2):159–61.
38. Riggs BL, Khosla S, Melton 3rd LJ. Sex steroids and the construction and conservation of the adult skeleton. Endocr Rev. 2002;23(3):279–302.
39. Prior JC, Vigna YM, Schechter MT, Burgess AE. Spinal bone loss and ovulatory disturbances. N Engl J Med. 1990;323(18):1221–7.
40. Snead DB, Weltman A, Weltman JY, Evans WS, Veldhuis JD, Varma MM, et al. Reproductive hormones and bone mineral density in women runners. J Appl Physiol. 1992;72(6):2149–56.
41. Cobb KL, Bachrach LK, Greendale G, Marcus R, Neer RM, Nieves J, et al. Disordered eating, menstrual irregularity, and bone mineral density in female runners. Med Sci Sports Exerc. 2003;35(5):711–9.

42. Christo K, Prabhakaran R, Lamparello B, Cord J, Miller KK, Goldstein MA, et al. Bone metabolism in adolescent athletes with amenorrhea, athletes with eumenorrhea, and control subjects. Pediatrics. 2008;121(6):1127–36.
43. Lewiecki EM, Gordon CM, Baim S, Leonard MB, Bishop NJ, Bianchi ML, et al. International Society for Clinical Densitometry 2007 Adult and Pediatric Official Positions. Bone. 2008; 43(6):1115–21.
44. Ducher G, Turner AI, Kukuljan S, Pantano KJ, Carlson JL, Williams NI, et al. Obstacles in the optimization of bone health outcomes in the female athlete triad. Sports Med. 2011;41(7):587–607.
45. Vescovi JD, Jamal SA, De Souza MJ. Strategies to reverse bone loss in women with functional hypothalamic amenorrhea: a systematic review of the literature. Osteoporos Int. 2008;19(4):465–78.
46. Cobb KL, Bachrach LK, Sowers M, Nieves J, Greendale GA, Kent KK, et al. The effect of oral contraceptives on bone mass and stress fractures in female runners. Med Sci Sports Exerc. 2007;39(9):1464–73.
47. Miller KK, Lawson EA, Mathur V, Wexler TL, Meenaghan E, Misra M, et al. Androgens in women with anorexia nervosa and normal-weight women with hypothalamic amenorrhea. J Clin Endocrinol Metab. 2007;92(4):1334–9.
48. Warren MP, Brooks-Gunn J, Fox RP, Holderness CC, Hyle EP, Hamilton WG, et al. Persistent osteopenia in ballet dancers with amenorrhea and delayed menarche despite hormone therapy: a longitudinal study. Fertil Steril. 2003;80(2):398–404.
49. Keen AD, Drinkwater BL. Irreversible bone loss in former amenorrheic athletes. Osteoporos Int. 1997;7(4):311–5.
50. Drinkwater BL, Bruemner B, Chesnut 3rd CH. Menstrual history as a determinant of current bone density in young athletes. JAMA. 1990;263(4):545–8.
51. Wiksten-Almstromer M, Hirschberg AL, Hagenfeldt K. Reduced bone mineral density in adult women diagnosed with menstrual disorders during adolescence. Acta Obstet Gynecol Scand. 2009;88(5):543–9.
52. Anderson TJ, Uehata A, Gerhard MD, Meredith IT, Knab S, Delagrange D, et al. Close relation of endothelial function in the human coronary and peripheral circulations. J Am Coll Cardiol. 1995;26(5):1235–41.
53. Neunteufl T, Katzenschlager R, Hassan A, Klaar U, Schwarzacher S, Glogar D, et al. Systemic endothelial dysfunction is related to the extent and severity of coronary artery disease. Atherosclerosis. 1997;129(1):111–8.
54. Zeni Hoch A, Dempsey RL, Carrera GF, Wilson CR, Chen EH, Barnabei VM, et al. Is there an association between athletic amenorrhea and endothelial cell dysfunction? Med Sci Sports Exerc. 2003;35(3):377–83.
55. Rickenlund A, Eriksson MJ, Schenck-Gustafsson K, Hirschberg AL. Amenorrhea in female athletes is associated with endothelial dysfunction and unfavorable lipid profile. J Clin Endocrinol Metab. 2005;90(3):1354–9.
56. Hoch AZ, Papanek P, Szabo A, Widlansky ME, Schimke JE, Gutterman DD. Association between the female athlete triad and endothelial dysfunction in dancers. Clin J Sport Med. 2011;21(2):119–25.
57. Halcox JP, Schenke WH, Zalos G, Mincemoyer R, Prasad A, Waclawiw MA, et al. Prognostic value of coronary vascular endothelial dysfunction. Circulation. 2002;106(6):653–8.
58. Targonski PV, Bonetti PO, Pumper GM, Higano ST, Holmes Jr DR, Lerman A. Coronary endothelial dysfunction is associated with an increased risk of cerebrovascular events. Circulation. 2003;107(22):2805–9.

Chapter 2
Nutritional Recommendations for the Young and Aging Females

Alexis M. Ziemba, Emily J. Curry, Jingyi Gong, and Elizabeth Matzkin

Abstract Food selection is highly dependent upon cultural practices and regional food availability. The United States Department of Agriculture has repeatedly made modifications to nutritional recommendations for the nation over the past century. The food pyramid was created in 1992 in an attempt to simplify key nutritional points. The food pyramid was revamped in 2005 when MyPyramid was created. Since then, MyPlate is the most widely used visual tool for nutritional education. In the young female, proper calcium and vitamin D intake is key in the prevention of osteoporosis later in life. Sarcopenia and osteoporosis are dual concerns for the aging female, since both increase the risk of fractures. Sarcopenia prevention can occur through proper nutrition, including sufficient protein intake. Exercise is also important for sarcopenia and osteoporosis prevention. Requirements for proper nutrition vary between patients and a patient specific approach is crucial to overall health in the young and aging female.

Keywords Nutrition • Bone mineral density • Rickets • Calcium • Vitamin D • Iron • Sarcopenia • Obesity • Essential amino acid • Leucine • Anabolic resistance syndrome

---

A.M. Ziemba, BA
Department of Anesthesia, Pain, and Critical Care,
Massachusetts General Hospital, Boston, MA 02114, USA
e-mail: aziemba@partners.org

E.J. Curry, BA • E. Matzkin, MS, MD (✉)
Department of Orthopaedic Surgery, Brigham and Women's
Hospital, 75 Francis Street, Boston, MA 02115, USA
e-mail: ecurry1@partners.org; ematzkin@partners.org

J. Gong, BS
Department of Molecular Cellular Biology/Biochemistry,
Brown University, 70 Ship Street, Providence, RI 02906, USA
e-mail: jingyi_gong@brown.edu

E. Mody, E. Matzkin (eds.), *Musculoskeletal Health in Women*,
DOI 10.1007/978-1-4471-4712-1_2,

Abbreviations

ATP	Adenosine triphosphate
BMI	Body mass index
EAAs	Essential amino acids
IL-1	Interleukin-1
IL-6	Interleukin-6
IU	International Unit
LBW	Low birth weight
RDA	Recommended Dietary Allowance
rhGH	Recombinant human growth hormone
TNF-α	Tumor necrosis factor alpha
USDA	United States Department of Agriculture
UVB	Ultraviolet-B

Introduction

Food choice is product of culture and geographical location. Many Asian countries consider rice a staple, while Mediterranean countries emphasize the consumption of olive-based products as a result of established regional cultivation practices [1, 2]. Coastal Asian regions and the Mediterranean obtain daily protein predominantly from seafood due to coastal proximity. In contrast, meats and sweets are consumed as infrequently as monthly due to limited availability and cultural practices.

In the United States, the United States Department of Agriculture (USDA) has frequently changed the recommendations for a balanced diet. The USDA developed the first recommendations for healthy eating as early as 1894 [3]. Since then, many efforts have been made to educate the United States population about sufficient nutrition. Most notably, the Food Guide Pyramid was established in 1992, which included five main food groups: meat, dairy, vegetables, fruit, and grain in addition to fats, oils, and sweets [3]. The USDA recommended a specific number of servings per food group; however, there were no measurements, whether in mass or volume, to define a serving.

MyPyramid (Fig. 2.1) was developed in 2005 for a more aesthetically pleasing model using vertical strips representing daily portions for the five main food groups described in 1992 [4]. Exercise and recommended daily calorie intakes based on age, gender and food volumes were included in this model, instead of servings [4]. Despite the remodeled "MyPyramid", Gao et al. suggested that by following either the 1992 or the 2005 food pyramid resulted in excessive calorie consumption leading to obesity [5]. However, they did indicate that "MyPyramid" was better equipped to supply the necessary nutrients compared to the 1992 pyramid suggesting progress was made with the 2005 recommendations [5].

In response to the inadequacies of the 1992 and 2005 food pyramid, the USDA shifted to MyPlate in 2011 (Fig. 2.2) [6]. The online nutrition site www.choosemyplate.gov provides comprehensive sections for each of the five main food groups, describes

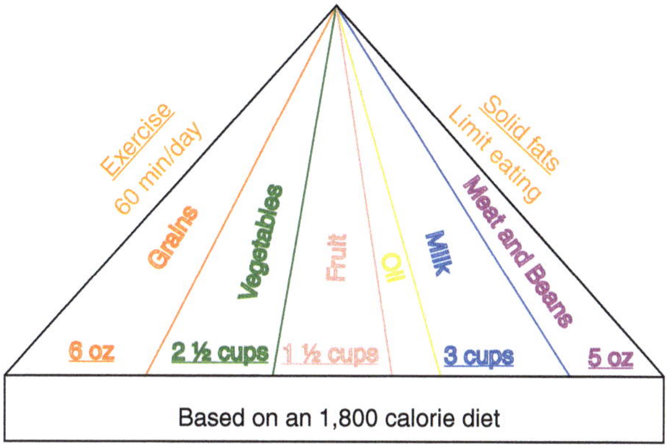

Fig. 2.1 MyPyramid was created by the USDA in 2005 to more effectively educate patients about proper nutrition (Inspired by U.S. Department of Agriculture [USDA]. 2005. MyPyramid. gov. Retrieved March 21, 2006, from http://www.mypyramid.gov. Created from data in Fowles [4])

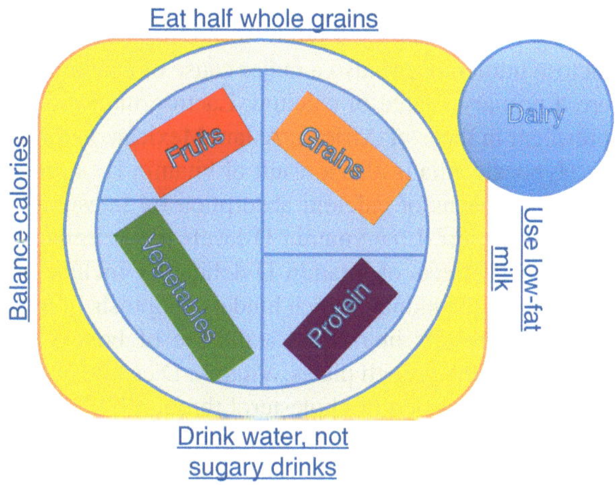

Fig. 2.2 MyPlate was created by the USDA in 2011 to address the inadequacies of MyPyramid and the Food Pyramid (*Source*: USDA [6])

foods qualifying for each food group, and specifically defines a "serving" for each food type. MyPlate also recommends how to choose healthy foods within a group, such as consuming 50 % whole grains instead of 100 % refined grains. MyPlate also reinforces the importance of a balanced diet and sufficient exercise to remain healthy.

Although there are recommendations to guide patients in nutritional and lifestyle choices, clinicians advising patients about proper dietary considerations must be aware that proper nutrition is patient specific and is affected by age and sex. This chapter aims to discuss the nutritional concerns for the growing and aging female and strategies to combat problems with bone accretion and muscle atrophy.

Bone and Growth: Nutrition for Developing Girls

The skeletal system contains roughly 99 % of the body's calcium [7]. Girls experience the maximum accumulation of calcium for bone growth around 12.5 years of age [8]. Girls subsequently develop 40 % of the bone mass necessary for their entire life span during the 3–4-year pubertal growth period [7]. There are numerous causes that prevent bone accretion and ultimately slow development, including food allergies, genetic mutations, and malnutrition.

Food allergies are becoming increasingly prevalent and can cause low calcium intake or absorption. People who have celiac disease, an inability to digest gluten, are often found to have a low bone mineral density as a result of hypocalcemia. Their physiological response to gluten involves increased cytokines, including IL-1, IL-6, and TNF-α, which cause inflammation and damage to the intestinal villi. This results in poor intestinal absorption of calcium and vitamin D. Low levels of calcium trigger the production of parathyroid hormone, which, in conjunction with the cytokines, acts to increase calcium absorption in the intestines and activates osteoclasts to break down bone to increase calcium levels. Enhanced osteoclast activity results in low bone mineral density and can lead to increased risk of fracture [9]. If children adhere to a gluten-free diet for approximately 1 year, bone mineral density is increased by a return-to-normal calcium absorption [10, 11]. Similarly, lactose intolerance is also a major cause of calcium deficiency since dairy products are a primary source of calcium. Lactose intolerance has been shown to be more prevalent in those of African, Asian, Mexican, and Native American descent [7].

Hypocalcemia is also a result of vitamin D (a pre-pro-hormone) deficiency [12, 13]. One means of calcium absorption by intestinal villi is vitamin D-dependent active transport. Low vitamin D levels have similar effects as low calcium levels. Additional effects of vitamin D deficiency include a decrease in phosphate reabsorption by the kidney, which hinders apoptosis of hypertrophic chondrocytes. This ultimately results in rickets, a disease of bone growth where mineralization is impaired at the growth plates. Vitamin D can be obtained either through the diet or from the conversion of cholesterol through UVB exposure. Those more at risk for vitamin D deficiency are children breast-fed by vitamin D-deficient mothers, those living in high altitudes (due to the angle of sun exposure) and greater distance from the equator, and those with greater melanin levels. Symptoms of rickets are often seen within 6–24 months and include abnormalities in bones, enlargement of bone junctions such as the wrist, and motor skill deficiencies [12, 14].

In addition to vitamin D dietary deficiency, calcium absorption can also be impaired regardless of vitamin D levels by polymorphisms of the vitamin D receptors or enzymes involved in vitamin D metabolism [7, 14, 15]. For example, the renal enzyme, 25(OH)D-1-α-hydroxylase, that produces the vitamin D metabolite, $1,25(OH)_2D$, has been targeted in the autosomal recessive disorder called vitamin D-dependent rickets type 1 [16]. Three mutations of 25(OH)D-1-α-hydroxylase have been identified that result in decreased $1,25(OH)_2D$ levels, which ultimately leads to rickets [16].

Bone development is also correlated with a child's body development and overall body weight. One study demonstrated that 18 % of children that did not drink cow's milk were overweight [17]. This could be partly attributed to the lack of calcium consumption, which is involved in lipolysis of adipose tissue. Therefore, the lack of dairy consumption simultaneously increases body fat and reduces calcium intake for bone accrual [17]. A high body weight has also been seen as a risk factor of distal forearm fractures in girls, possibly due to greater adipose composition [18]. Conversely, a study that measured a number of bone characteristics related to bone thickness, bone strength, and surrounding muscle and fat composition in children ages 7–10 found overweight children to have a greater bone strength than normal weight children [19]. The results suggest the greater bone strength can be attributed, in part, to greater muscle mass.

Thus, it can also be inferred that children born with a low birth weight (LBW) may have weaker bones due to a low muscle mass. LBW can be associated with malnutrition and stunted growth along with other health problems during adulthood, such as hypertension. Around the world, women suffer from malnutrition for economic and educational reasons. Vitamin A, iron, iodine, and/or zinc deficiencies in women during pregnancy can result in intrauterine growth restriction and/or preterm delivery [20]. Vitamin A deficiency can be due to diets void of dairy, eggs, and fruits and vegetables [20]. Physiological iodine shortage can be attributed to living in areas that have low traces of iodine in the soil and water [20]. As with calcium, zinc insufficiency can also be caused by eating an abundance of phytates, which bind to minerals and prevent absorption into the body [20, 21]. One dietary cause of anemia in children and pregnant mothers is a lack of fish, meats, and poultry in the diet, combined with a high intake of cereals and legumes [22].

Nutritional Suggestions for Growing Girls

Children ages 9–19 should have a daily calcium intake of 1,300 mg [13]. One study examining pubertal stage 2 girls found that they consumed an average of 830 mg calcium/day [23]. Those whose diets were supplemented with 670 mg calcium citrate/day, a calcium salt, had greater bone mineralization during their growth spurt [23]. For infants, human milk is superior to formula due to greater calcium bioavailability [7]. For children ages 4–8, consuming three 8-oz glasses of milk is sufficient for daily calcium intake, while four 10-oz glasses is adequate for pubescent children [7]. Calcium carbonate and calcium citrate can be given as calcium supplements [13]. Calcium citrate was demonstrated to have greater bioavailability when taken in 500 mg doses by postmenopausal women [24]. Excessive calcium intake also poses risks, including kidneys stones, poor absorption of other divalent cations (such as magnesium), and kidney failure [13, 25].

For lactose intolerant individuals, fermented dairy products, such as yogurt and hard cheese, may be consumed [7]. Obtaining enough calcium is more difficult for weaned children on a vegan diet. Nondairy foods high in calcium include soy products, leafy vegetables, and cereals [7, 21]. However, the bioavailability of soy products is low due to the presence of phytates, phosphorus-containing compounds that chelate ions, including zinc, making them difficult to absorb. Other substances that make calcium absorption difficult include excessive intake of protein, caffeine, or alcohol [7].

Exposure to sunlight and 5,000–15,000 IU/day vitamin D is recommended for 4–8 weeks to treat rickets and vitamin D deficiency [12]. As there are two types of vitamin D: D_2 or ergocalciferol and D_3 or cholecalciferol, there is debate about which is more effective. One study demonstrated that administration of both types (1,000 IU/day) for 11 weeks resulted in similar amounts of 25(OH)D, a measure of vitamin D; however, a second study that administered 50,000 IU/week of both D_2 and D_3 saw better sustenance of 25(OH)D levels with D3 consumption, potentially due to the shorter half-life of ergocalciferol [26–28]. Results imply that daily administration of either vitamin D_2 or D_3 consistently maintains necessary levels of vitamin D.

A study was conducted on children in India that were prepubertal and growth hormone deficient. After 1 year of observation, these children were given daily treatments of recombinant human growth hormone (rhGH, 20 mg/m^2/week), rhGH, calcium (500 mg), monthly vitamin D_3 (60,000 IU), or rhGH, calcium, zinc (determined by recommended dietary allowance), and monthly vitamin D_3. Through the measurement of biochemical parameters and bone mineral density, the subjects had greater bone mineralization with growth hormone therapy and was further increased by supplementation with calcium, vitamin D, and zinc [29].

Low birth weight due to maternal malnourishment is another cause for concern and may be avoided by protein supplementation (700 kcal/day) [20]. Certainly, amino acids essential for protein synthesis can be obtained in meat, poultry, and fish. As vegetarian diets are becoming more prominent, additional sources of amino acids are needed. Many vegetables contain the 20 essential amino acids; however, they are more difficult for humans to utilize through digestion. Legumes, nuts, and seeds are sources that contain plentiful amounts of protein [21]. Zinc deficiency is also more common for vegetarians because 50 % of zinc is consumed through animal protein. As with amino acids, zinc intake can be supplemented by eating legumes and soy products [21]. There was a study of children with anemia in Mexican villages, age 6–42 months, who were given a variation of iron treatment to remedy the problem. The treatments included an iron supplement, an iron and folic acid supplement, a multiple micronutrient supplement, a micronutrient-fortified complementary food as porridge powder, or zinc, iron, and ascorbic acid-fortified water. The efficacy of each treatment was monitored by measuring height, weight, and biochemical measurements including hemoglobin, iron, ferritin, and C-reactive protein concentrations. Increased concentrations of hemoglobin and total iron were found with each treatment; however, the iron and folic acid and multiple micronutrient supplements were the most effective, suggesting the supplements in addition to iron are important in remedying anemia. The micronutrient-fortified complementary food was preferred over the supplements [22]. Iron absorption can be enhanced by vegetable consumption that contains high levels of vitamin C [21].

Sarcopenia: Nutrition for Maintaining Muscle Mass in Elderly Women

Sarcopenia is prevalent in the elderly population and is characterized by a loss of muscle mass and strength. Atrophy resulting from sarcopenia decreases mobility and increases frailty of older adults and ultimately increasing the risk of falls [30, 31].

Elderly patients were also found to be more susceptible to sarcopenia due to greater periods of hospitalization. Confinement to a hospital bed results in less use of muscles and atrophy of those muscles. A study of hospitalized patients ages 65 and older were found that sarcopenia correlated with older patients and longer hospital stays [32]. Conversely, it has also been shown that the majority of muscle loss happens within the first few days [33]. Increased protein ubiquitination, which indicates proteins are being tagged for destruction, was seen 48 h after limb immobilization, suggesting muscle catabolism begins almost immediately following muscle disuse [33].

Sarcopenia is also seen in elderly adults living independently. Older adults living in North Carolina were tested on tasks, including opening jars, lifting full glasses, using a can opener, opening cartons, and open plastic bags. The difficulty in these activities due to poor mobility makes meal preparation and consumption difficult and thus results in poor nutrition. Elderly adults that had greater difficulty with the tasks were found to have lower levels of calcium, vitamin D, magnesium, and phosphorus, essential for the maintenance of the musculoskeletal system [34].

Malnutrition is common in older adults due to problems swallowing or chewing as a result of mouth disease, gum disease, or poor oral hygiene. A study of Dutch patients ages 65 and older receiving home care suggests that malnutrition due to low protein intake is a cause of sarcopenia. The patients that had reduced mobility and malnutrition had a greater risk of sarcopenia and they suggested that this could in part be remedied by better nutrition [30].

Malnutrition can also be a result of muscle disuse from events such as hospitalization accompanied by illnesses and/or injuries that decrease appetite. One study evaluated hospitalized patients with three different nutritional screening tests and found that more than 55 % of the patients suffered from malnutrition [35]. A study conducted in the Netherlands on community-dwelling, frail, and institutionalized elderly revealed that the protein intake at breakfast in all three groups was lower than the recommended amount. The daily protein intake for the institutionalized was 35 % lower than the suggested average [36].

In addition to muscle disuse and malnutrition, sarcopenia is a comorbidity of obesity. Obesity and sarcopenia are two interrelated conditions that affect health in the elderly. The unfavorable fat-to-muscle ratio seen in patients with sarcopenia is exacerbated when a patient becomes obese [37]. Sarcopenia was found to be significantly associated with sarcopenic obesity, a body composition condition in which obesity is masked by increase in weight as the concomitant decrease in lean mass cancels out the increase in fat mass. Body mass index (BMI) is usually used in defining obesity [38]. While increased BMI is usually associated with increased body fat, it measures body weight accounting for height and does not provide information regarding the fat-to-muscle ratio. It is therefore possible that co-occurrence of increased body fat and decreased muscle mass in patients with sarcopenic obesity increases health risks without significant increase in BMI [39]. Therefore, BMI is not always indicative of health risks associated with obesity and focusing on body composition; rather, BMI can be helpful in assessing individuals with sarcopenic obesity in older ages [39, 40].

At the cellular level, perturbations of protein turnover have a significant effect on sarcopenia. Studies suggest there is a downregulation of mitochondrial pathways involved in protein synthesis and upregulation of genes implicated in protein degradation by 14 days following limb immobilization [33]. Increases in the rate of protein breakdown and decreases in the rate of protein synthesis contribute to the loss of muscle mass in sarcopenia. Generally, there is a daily protein turnover rate of 1–2 % [41]. While the primary effect of muscle disuse is loss of muscle mass, bone mineral density also decreases due to muscle disuse. Unilateral limb suspension resulted in decreases in the bone mineral density of the distal tibia epiphysis [42]. Losses in calcium result in increased bone frailty in addition to muscle mass and strength. Men have significantly higher skeletal muscle mass and bone mineral density compared to women, so this result is more profound in women [43, 44].

Nutritional Suggestions for the Elderly

A high protein diet has been implicated in better overall health in elderly patients. An abundance of protein enables protein synthesis, which is essential for building muscle mass. Greater muscle mass allows for greater strength, which ultimately decreases the risk of falls and allows the elderly an increased quality of life, including easier meal preparation and increased mobility to socialize with friends [45].

Studies advocate various forms of protein supplementation for skeletal muscle maintenance. A decrease in whole-body protein synthesis can be seen in as little as 7 days of bed rest [46]. With consumption of 0.6–1.0 g/kg protein (from caseinates and soy protein), improvement was seen in leucine metabolism kinetics to levels that were similar to the non-bed rest control [46]. Furthermore, a study by Campbell et al. showed a gradual decline in urinary nitrogen excretion and decreased mid-thigh area in older individuals consuiming the recommended dietary allowance (RDA) for protein (0.8 g protein/kg/day) [39, 47]. This finding suggests that more protein should be ingested by healthy older individuals. Maintaining an intake of 25–30 g protein per meal has been shown to most effectively promote muscle protein synthesis in older ages and therefore can be recommended for sarcopenia prevention in the elderly [39, 48].

Results have shown that protein sources that are easier to digest and those with higher leucine content are more effective at initiating de novo protein synthesis in muscle [49–52]. Hydrolyzed casein has been shown to be more effective than intact casein at increasing postprandial amino acid concentration in elderly men, and whey protein, which has a high leucine content and is even more readily digestible, was the most superior in triggering protein synthesis in elderly men [49, 50]. Unfortunately, the same conclusions cannot be generalized to women, because the literature is lacking in studies looking specifically at findings for women. However, studies have shown that diets supplemented with greater levels of leucine (0.052 g/kg) result in greater plasma leucine concentration and fractional synthesis rate 5 h after eating [51]. There is also evidence that the infusion of leucine increases the retention of amino acids through protein turnover modulation by inhibiting protein catabolism [52].

Other studies infer that increased essential amino acid consumption may not be sufficient in the elderly population. Dardevet et al. discussed the idea of the anabolic threshold concept which states that as a side effect of aging, the body loses sensitivity to the hormones and amino acids that are able to initiate protein anabolism and thus requires greater amounts [53]. Consuming subthreshold levels results in muscle atrophy. Despite many studies suggesting the intake of more leucine to be sufficient, the anabolic increases are transient and may not have an overall effect on muscle atrophy. Thus, Arnal et al. have employed the "protein pulse feeding" method which involves the consumption of 80 % of the recommended daily protein to be consumed in a single meal; this results in a prolonged anabolic response and ultimately prevents atrophy [54]. A study of women over the age of 68 saw greater protein retention in women who consumed the protein in a protein pulse manner for 14 days [54].

Consuming too much protein to reach the anabolic threshold can cause additional problems, particularly in elderly patients who often have renal problems. Thus, the restoration of the normal anabolic threshold by reducing inflammation and increasing cytokine presence (i.e., IL-6) that are associated with increased muscle loss would be a more ideal approach [53]. One possibility for returning the anabolic stimuli threshold to normal levels is through physical activity [53].

In addition to various forms of protein consumption, other studies advise supplementing diets with creatine. In one study, creatine (5 g, four times daily) prevented muscle atrophy and weakness compared to placebo control in young men immobilized in a cast [55]. This study should be considered with caution because the female population was not assessed for the overall use of creatine in the prevention of muscle atrophy. However, overall, the protective properties of creatine could be attributed to increases in phosphocreatine, which is hydrolyzed to form ATP in skeletal muscle, in addition to increasing glycogen storage [41, 56].

To combat a general lack of energy intake in the elderly due to one of the many possible reasons for low consumption, snacking is important. It has been found that patients aged 60 and older who snacked more frequently (more than four times per day) and acquired 20–30 % of their daily energy from snacking had a faster gait [57]. However, trying to compensate for muscle loss by overeating without discrimination about food type will not prevent or reverse muscle atrophy; but instead, as would be expected, will lead to obesity. [41] For sarcopenic obesity management, energy restriction alone could lead to reduction in both fat and muscle mass, promoting an unfavorable ratio of body fat to muscle. Instead, daily energy intake should be reduced by no more than 956 kcal/day [37].

Supplementation of high-quality protein has also been considered in managing sarcopenic obesity [39]. Both whey protein and essential amino acids (EAAs) have been shown to promote weight loss in obese elderly individuals with modest loss of muscle mass [58]. EAAs have been suggested to positively affect protein synthesis through the AKT/mTOR signaling pathway [59]. Resistance exercise followed by ingestion of EAAs, especially leucine, with carbohydrate (EAA+CHO) can stimulate the signaling pathway [60].

A study looking at adults over the age of 60 (BMI 17–25) found that those who had a higher daily consumption of calcium had a lower incidence of sarcopenia [61].

Table 2.1 Recommended calcium intake

Sex	Age (years)	Calcium recommended daily allowances (mg)
Male and female	9–18	1,300
Male and female	19–50	1,000
Female	51–70	1,200
Male	51–70	1,000
Male and female	>70	1,200
Female – pregnant and breast-feeding	Teen	3,000
Female – pregnant and breast-feeding	Adult	2,500

Adapted with permission from Ross et al. [66]

Table 2.2 Recommended protein intake for the elderly

Form of protein	General [47]	Essential amino acids [48]	Leucine [51]
Daily intake	0.8 g/kg	16.5 g ×5	0.052 g/kg

Previous studies have shown that high calcium intake results in a lower adipose content [62]. Greater daily calcium intake could control sarcopenia in non-obese patients, but also obese patients by both lowering body fat (stopping the permeation of fat to skeletal muscle) and preventing the loss of muscle mass [61].

Due to its role in calcium absorption, low levels of vitamin D have also been shown to impair mobility. One study examined the levels of 25-hydroxyvitamin D in subjects ages 60 and older [63]. When doing an 8-ft walk test and sit-to-stand test, patients with 25-hydroxyvitamin levels greater than 40 nmol/L completed the tasks faster and thus had better musculoskeletal function [63]. Furthermore, in a study of women ages 63 and older, the women were given 1,200 mg calcium and 800 IU vitamin D in the form of cholecalciferol per day. Over a 3-month period, these women saw a 49 % reduction in falls compared to before diet supplementation [64].

Conclusions

Bone accretion and muscle maintenance are essential for mobility of growing girls and aging women, respectively. Adequate nutrition in young girls can be impacted by allergies, genetic mutations, and their pregnant mother's diet. The developing child requires a diet abundant in calcium and vitamin D to increase bone mineral density (Table 2.1) [65]. Low consumption of either can result in the bone dysmorphia disease and rickets and increases the risk of osteoporosis later in life.

There are many health and environmental factors that result in muscle atrophy in elderly women. Studies suggest the consumption of foods high in amino acids promote protein synthesis, but caution is necessary to avoid too much fat intake in the process (Table 2.2). Furthermore, leucine has been shown to be the most effective in modulating protein turnover rates because it not only promotes protein synthesis but also has been shown to inhibit catabolism. In both cases, proper nutrition should be accompanied by physical activity, which is necessary for musculoskeletal maintenance and continued mobility.

References

1. Asian Diet Pyramid. In: OldWays. Edited; 2013. http://oldwayspt.org/resources/heritage-pyramids/asian-diet-pyramid/overview.
2. Mediterranean Diet Pyramid. Edited, OldWays; 2013. http://oldwayspt.org/resources/heritage-pyramids/mediterranean-diet-pyramid/traditional-med-diet.
3. Davis C, Saltos E. Dietary recommendations and how they have changed over time. In: Frazao, E, editor. America's eating habits: changes and consequences. Washington, D.C.: Economic Research Service, United States Dept of Agriculture; 1999. Agriculture Information Bulletin No. 750. p. 33–50.
4. Fowles ER. What's a pregnant woman to eat? A review of current USDA Dietary Guidelines and MyPyramid. J Perinat Educ. 2006;15(4):28–33.
5. Gao X, Wilde PE, Lichtenstein AH, Tucker KL. The 2005 USDA Food Guide Pyramid is associated with more adequate nutrient intakes within energy constraints than the 1992 Pyramid. J Nutr. 2006;136(5):1341–6.
6. Agriculture, U. S. D. o.: MyPlate. In ChooseMyPlate.gov Website. Edited, Washington, D.C.; 2013. http://www.choosemyplate.gov.
7. Greer FR, Krebs NF, American Academy of Pediatrics Committee on Nutrition. Optimizing bone health and calcium intakes of infants, children, and adolescents. Pediatrics. 2006;117(2):578–85.
8. Bailey DA, Martin AD, McKay HA, Whiting S, Mirwald R. Calcium accretion in girls and boys during puberty: a longitudinal analysis. J Bone Miner Res. 2000;15(11):2245–50.
9. Goulding A, Cannan R, Williams SM, Gold EJ, Taylor RW, Lewis-Barned NJ. Bone mineral density in girls with forearm fractures. J Bone Miner Res. 1998;13(1):143–8.
10. Tau C, Mautalen C, De Rosa S, Roca A, Valenzuela X. Bone mineral density in children with celiac disease. Effect of a Gluten-free diet. Eur J Clin Nutr. 2006;60(3):358–63.
11. Capriles VD, Martini LA, Areas JA. Metabolic osteopathy in celiac disease: importance of a gluten-free diet. Nutr Rev. 2009;67(10):599–606.
12. Pettifor JM, Prentice AP. The role of vitamin D in paediatric bone health. Best Pract Res Clin Endocrinol Metab. 2011;24:573–84.
13. Straub DA. Calcium supplementation in clinical practice: a review of forms, doses, and indications. Nutr Clin Pract. 2007;22(3):286–96.
14. Hazell TJ, DeGuire JR, Weiler HA. Vitamin D: an overview of its role in skeletal muscle physiology in children and adolescents. Nutr Rev. 2012;70(9):520–33.
15. Schuster GL. Nutrients and gene expression. In: Kaput JR, Rodriguz L, editors. Nutritional genomics: discovering the path to personalized nutrition. Hoboken: Wiley; 2006. p. 168–9.
16. Kim CJ. Vitamin D dependent rickets type I. Korean J Pediatr. 2011;54(2):51–4.
17. Black RE, Williams SM, Jones IE, Goulding A. Children who avoid drinking cow milk have low dietary calcium intakes and poor bone health. Am J Clin Nutr. 2002;76(3):675–80.
18. Goulding A, Jones IE, Taylor RW, Manning PJ, Williams SM. More broken bones: a 4-year double cohort study of young girls with and without distal forearm fractures. J Bone Miner Res. 2000;15(10):2011–8.
19. Ducher G, Bass SL, Naughton GA, Eser P, Telford RD, Daly RM. Overweight children have a greater proportion of fat mass relative to muscle mass in the upper limbs than in the lower limbs: implications for bone strength at the distal forearm. Am J Clin Nutr. 2009;90(4):1104–11.

20. Ahmed T, Hossain M, Sanin KI. Global burden of maternal and child undernutrition and micronutrient deficiencies. Ann Nutr Metab. 2012;61 Suppl 1:8–17.
21. Amit M. Vegetarian diets in children and adolescents. Paediatr Child Health. 2010;15(5): 303–14.
22. Rosado JL, Gonzalez KE, Caamano Mdel C, Garcia OP, Preciado R, Odio M. Efficacy of different strategies to treat anemia in children: a randomized clinical trial. Nutr J. 2010;9:40.
23. Matkovic V, et al. Calcium supplementation and bone mineral density in females from childhood to young adulthood: a randomized controlled trial. Am J Clin Nutr. 2005;81(1): 175–88.
24. Heller HJ, Greer LG, Haynes SD, Poindexter JR, Pak CY. Pharmacokinetic and pharmacodynamic comparison of two calcium supplements in postmenopausal women. J Clin Pharmacol. 2000;40(11):1237–44.
25. Ashley RH, Williams AJ. Divalent cation activation and inhibition of single calcium release channels from sheep cardiac sarcoplasmic reticulum. J Gen Physiol. 1990;95(5):981–1005.
26. Holick MF, et al. Vitamin D2 is as effective as vitamin D3 in maintaining circulating concentrations of 25-hydroxyvitamin D. J Clin Endocrinol Metab. 2008;93(3):677–81.
27. Heaney RP, Recker RR, Grote J, Horst RL, Armas LA. Vitamin D(3) is more potent than vitamin D(2) in humans. J Clin Endocrinol Metab. 2011;96(3):E447–52.
28. Romagnoli E, Mascia ML, Cipriani C, Fassino V, Mazzei F, D'Erasmo E, Carnevale V, Scillitani A, Minisola S. Short and long-term variations in serum calciotropic hormones after a single very large dose of ergocalciferol (vitamin D2) or cholecalciferol (vitamin D3) in the elderly. J Clin Endocrinol Metab. 2008;93(8):3015–20.
29. Ekbote V, Khadilkar A, Chiplonkar S, Mughal Z, Khadilkar V. Enhanced effect of zinc and calcium supplementation on bone status in growth hormone-deficient children treated with growth hormone: a pilot randomized controlled trial. Endocrine. 2013;43(3):686–95.
30. Meijers JM, Halfens RJ, Neyens JC, Luiking YC, Verlaan G, Schols JM. Predicting falls in elderly receiving home care: the role of malnutrition and impaired mobility. J Nutr Health Aging. 2012;16(7):654–8.
31. Arango-Lopera VE, Arroyo P, Gutierrez-Robledo LM, Perez-Zepeda MU, Cesari M. Mortality as an adverse outcome of sarcopenia. J Nutr Health Aging. 2013;17(3):259–62.
32. Gariballa S, Alessa A. Sarcopenia: prevalence and prognostic significance in hospitalized patients. Clin Nutr. 2013 Jan 30. pii: S0261-5614(13)00036-8. doi:10.1016/j.clnu.2013.01.010. [Epub ahead of print].
33. Abadi A, et al. Limb immobilization induces a coordinate down-regulation of mitochondrial and other metabolic pathways in men and women. PLoS One. 2009;4(8):e6518.
34. Sharkey J, Johnson CM, Dean WR. Physical limitations in meal preparation and consumption are associated with lower musculoskeletal nutrient (calcium, vitamin D, magnesium, and phosphorus) intakes in homebound older adults. J Nutr Health Aging. 2012;16(8):675–7.
35. Vanis N, Mesihovic R. Application of nutritional screening tests for determining prevalence of hospital malnutrition. Med Arh. 2008;62(4):211–4.
36. Tieland M, Borgonjen-Van den Berg KJ, van Loon LJ, de Groot LC. Dietary protein intake in community-dwelling, frail, and institutionalized elderly people: scope for improvement. Eur J Nutr. 2012;51(2):173–9.
37. Parr EB, Coffey VG, Hawley JA. 'Sarcobesity': a metabolic conundrum. Maturitas. 2013;74(2): 109–13.
38. Silva AO, Karnikowski MG, Funghetto SS, Stival MM, Lima RM, de Souza JC, Navalta JW, Prestes J. Association of body composition with sarcopenic obesity in elderly women. Int J Gen Med. 2013;6:25–9.
39. Li Z, Heber D. Sarcopenic obesity in the elderly and strategies for weight management. Nutr Rev. 2012;70(1):57–64.
40. Zamboni M, Mazzali G, Zoico E, Harris TB, Meigs JB, Di Francesco V, Fantin F, Bissoli L, Bosello O. Health consequences of obesity in the elderly: a review of four unresolved questions. Int J Obes (Lond). 2005;29(9):1011–29.

41. Wall BT, van Loon LJ. Nutritional strategies to attenuate muscle disuse atrophy. Nutr Rev. 2013;71(4):195–208.
42. Rittweger J, Winwood K, Seynnes O, de Boer M, Wilks D, Lea R, Rennie M, Narici M. Bone loss from the human distal tibia epiphysis during 24 days of unilateral lower limb suspension. J Physiol. 2006;577(Pt 1):331–7.
43. Janssen I, Heymsfield SB, Wang ZM, Ross R. Skeletal muscle mass and distribution in 468 men and women aged 18–88 yr. J Appl Physiol. 2000;89(1):81–8.
44. Looker AC, Melton 3rd LJ, Harris T, Borrud L, Shepherd J, McGowan J. Age, gender, and race/ethnic differences in total body and subregional bone density. Osteoporos Int. 2009; 20(7):1141–9.
45. Wolfe RR. The role of dietary protein in optimizing muscle mass, function and health outcomes in older individuals. Br J Nutr. 2012;108 Suppl 2:S88–93.
46. Stuart CA, Shangraw RE, Peters EJ, Wolfe RR. Effect of dietary protein on bed-rest-related changes in whole-body-protein synthesis. Am J Clin Nutr. 1990;52(3):509–14.
47. Campbell WW, Trappe TA, Wolfe RR, Evans WJ. The recommended dietary allowance for protein may not be adequate for older people to maintain skeletal muscle. J Gerontol A Biol Sci Med Sci. 2001;56(6):M373–80.
48. Paddon-Jones D, Rasmussen BB. Dietary protein recommendations and the prevention of sarcopenia. Curr Opin Clin Nutr Metab Care. 2009;12(1):86–90.
49. Koopman R, et al. Ingestion of a protein hydrolysate is accompanied by an accelerated in vivo digestion and absorption rate when compared with its intact protein. Am J Clin Nutr. 2009;90(1): 106–15.
50. Pennings B, Boirie Y, Senden JM, Gijsen AP, Kuipers H, van Loon LJ. Whey protein stimulates postprandial muscle protein accretion more effectively than do casein and casein hydrolysate in older men. Am J Clin Nutr. 2011;93(5):997–1005.
51. Rieu I, Balage M, Sornet C, Giraudet C, Pujos E, Grizard J, Mosoni L, Dardevet D. Leucine supplementation improves muscle protein synthesis in elderly men independently of hyperaminoacidaemia. J Physiol. 2006;575(Pt 1):305–15.
52. Nair KS, Schwartz RG, Welle S. Leucine as a regulator of whole body and skeletal muscle protein metabolism in humans. Am J Physiol. 1992;263(5 Pt 1):E928–34.
53. Dardevet D, Remond D, Peyron MA, Papet I, Savary-Auzeloux I, Mosoni L. Muscle wasting and resistance of muscle anabolism: the "anabolic threshold concept" for adapted nutritional strategies during sarcopenia. Scientific World Journal. 2012;2012:269531.
54. Arnal MA, et al. Protein pulse feeding improves protein retention in elderly women. Am J Clin Nutr. 1999;69(6):1202–8.
55. Johnston AP, Burke DG, MacNeil LG, Candow DG. Effect of creatine supplementation during cast-induced immobilization on the preservation of muscle mass, strength, and endurance. J Strength Cond Res. 2009;23(1):116–20.
56. van Loon LJ, Murphy R, Oosterlaar AM, Cameron-Smith D, Hargreaves M, Wagenmakers AJ, Snow R. Creatine supplementation increases glycogen storage but not GLUT-4 expression in human skeletal muscle. Clin Sci (Lond). 2004;106(1):99–106.
57. Xu B, Yu GP, Zizza CA, Liu H, Zhao L. Snacking may improve physical function among older Americans. J Nutr Health Aging. 2013;17(4):393–7.
58. Coker RH, Miller S, Schutzler S, Deutz N, Wolfe RR. Whey protein and essential amino acids promote the reduction of adipose tissue and increased muscle protein synthesis during caloric restriction-induced weight loss in elderly, obese individuals. Nutr J. 2012;11:105.
59. Coker RH, Wolfe RR. Bedrest and sarcopenia. Curr Opin Clin Nutr Metab Care. 2012;15(1):7–11.
60. Dreyer HC, Drummond MJ, Pennings B, Fujita S, Glynn EL, Chinkes DL, Dhanani S, Volpi E, Rasmussen BB. Leucine-enriched essential amino acid and carbohydrate ingestion following resistance exercise enhances mTOR signaling and protein synthesis in human muscle. Am J Physiol Endocrinol Metab. 2008;294(2):E392–400.
61. Seo MH, Kim MK, Park SE, Rhee EJ, Park CY, Lee WY, Baek KH, Song KH, Kang MI, Oh KW. The association between daily calcium intake and sarcopenia in older, non-obese Korean

adults: the fourth Korea national health and nutrition examination survey (KNHANES IV) 2009. Endocr J. 2013;60(5):679–86.
62. Rosenblum JL, Castro VM, Moore CE, Kaplan LM. Calcium and vitamin D supplementation is associated with decreased abdominal visceral adipose tissue in overweight and obese adults. Am J Clin Nutr. 2012;95(1):101–8.
63. Bischoff-Ferrari HA, Dietrich T, Orav EJ, Hu FB, Zhang Y, Karlson EW, Dawson-Hughes B. Higher 25-hydroxyvitamin D concentrations are associated with better lower-extremity function in both active and inactive persons aged>or =60 y. Am J Clin Nutr. 2004;80(3):752–8.
64. Bischoff HA, et al. Effects of vitamin D and calcium supplementation on falls: a randomized controlled trial. J Bone Miner Res. 2003;18(2):343–51.
65. National Academy of Sciences. Tables S-1 and S-2 Dietary reference intakes for calcium and vitamin D. Washington, D.C.: National Academies Press; 2011.
66. Ross CA, Manson JE, Abrams SA, et al. The 2011 report on dietary reference intakes for calcium and vitamin D from the Institute of Medicine: what clinicians need to know. J Clin Endocrinol Metab. 2011;96:53–8.

Chapter 3
Women's Bone Health: Breathing Life into the Skeleton

Margaret Seton

Abstract The shape and size of bones define our physical selves and provide protection for internal organs and leverage sites for muscles. These mechanical properties of the skeleton create a static image of bone health, with impaired microarchitecture of bone and consequent fracture hallmarks of tissue failure. Yet bone is constantly created, modeled, and remodeled throughout the course of a woman's life, and its structure and health are intimately involved with calcium metabolism and sustenance of bone marrow cells. This chapter will discuss the systemic and local factors that affect bone health throughout the life of a woman, emphasizing the vulnerable, living quality of bone that is often lost in translation when tests are interpreted and treatments recommended.

Keywords Bone remodeling • Bone growth • Vitamin D • Osteoporosis • Fracture • Calcium • Bone health

Introduction

The shape and size of bones define our physical selves and provide protection for internal organs and leverage sites for muscles. These mechanical properties of the skeleton create a static image of bone health, with impaired microarchitecture of bone and consequent fracture hallmarks of tissue failure. Yet bone is constantly created, modeled, and remodeled throughout the course of a woman's life, and its structure and health are intimately involved with calcium metabolism and sustenance of bone marrow cells. This chapter will discuss the systemic and local factors that affect bone health throughout the life of a woman, emphasizing the vulnerable,

M. Seton, MD
Department of Rheumatology, Immunology, and Allergy,
Brigham and Women's Hospital, 75 Francis Street,
Boston, MA 02115, USA
e-mail: mseton@partners.org

E. Mody, E. Matzkin (eds.), *Musculoskeletal Health in Women*,
DOI 10.1007/978-1-4471-4712-1_3,

living quality of bone that is often lost in translation when tests are interpreted and treatments recommended. It is written with the intent to provide an overview of bone health in women, and to highlight current controversies.

Fetal Growth and Development

The fetal skeleton is formed predominantly by a process called endochondral bone formation, in which an anlage of the skeleton is laid down by chondrocytes, before osteoblasts are recruited to replace this with bone. Of the 206 bones formed by the skeleton, each bone is distinct, defined by its location and function and whether it informs the shape of the right or the left. Signaling in the growth plate, vascular invasion, and chondrocyte alignment all affect this process. Movement of the fetus is critical in defining bone shape and strength. By 9 weeks, the skeletal patterning has occurred [1]. Errors in the quality of collagen deposition (osteogenesis imperfecta), the timing and efficacy of mineralization (hypophosphatasia), and the definition of skeletal shape by chondrocytes (achondroplasia) result in impaired skeletal strength and function for the life of the individual. Many of these mutations are sporadic; some are inherited as autosomal dominant, some as recessive. Some of these mutations are associated with advanced paternal age.

During the 2nd and 3rd trimesters of pregnancy, adequate growth and mineralization of bone is required, and dramatic changes in 1,25-dihydroxyvitamin D (calcitriol) occur in the maternal circulation to effect this [2]. Maternal vitamin D insufficiency has been linked to increased risk for intrauterine growth retardation, preeclampsia, gestational diabetes, and higher rates of caesarian section [3]. How much vitamin D and in which populations were questions explored in a randomized clinical trial of supplementation in an endemically vitamin D-deficient Arab population. The findings from this study argue that vitamin D3 4,000 IU daily is safe and effective in ensuring adequate maternal vitamin D without adverse events [4]. While vitamin D sufficiency seems essential in fetal health, controversy exists whether maternal vitamin D insufficiency has long-lasting effects on skeletal health [5–8].

Screening by amniocentesis and imaging by ultrasound can begin to detect musculoskeletal abnormalities in utero. The ethical argument about how best to manage these findings is being raised by medical and laypersons across the country. Does early termination of pregnancies marked by Down's, achondroplasia, or osteogenesis imperfecta deprive a population of diversity and the human obligation of kindness [9]? Many who inherit these diseases struggle for identity and accommodation.

Childhood and Adolescence

The median age of menarche in the USA is 12.5 years, with 90 % of adolescent girls beginning their menses by age 13 [10]. Why this is critical to bone health has to do with the determinants of peak bone mass, which include normal puberty, regular menses, normal weight, and adequate nutrition. Children grow comparably during early childhood, but with puberty, girls with robust bones will grow differently from those with slender bones. Girls with small build tend to have endocortical infilling by age 8, 3 years earlier than their large-boned counterparts, and tend to have more porous bones in which there is cellular suppression of remodeling [11]. Not only does structure of these bones convey different thresholds for enduring extreme stress, such as military training, but the bone composition seems to carry different cellular physiology as well.

As systems biology is sorting out the consequences of altered cellular physiology on the structural integrity of bone [12], there are some clarion issues during adolescence in terms of skeletal health. This is when most growth – hence bone modeling – occurs over the 3 years or so before epiphyseal closure. Normal menstruation, estrogen levels, and nutrition are critical at this time. The bone mass accrual that occurs after puberty reflects mostly cortical bone apposition. Once peak bone mass is achieved by the mid-20s to early 30s, this will define a set point in a woman's life. Just as in fetal health, there is this window for skeletal patterning, so in adolescent health there is a window for bone growth and accrual of mass. Anorexia nervosa [13, 14], the female athlete triad [15], and drugs such as Depo-Provera [16] may result in lasting consequences to the skeleton, and teenagers should be screened for these issues. Adolescents should be counseled to receive adequate calcium, vitamin D, and caloric intake, avoid alcohol, and resist cigarette smoking.

Because of the changing size and shape of bones, bone mineral density examinations should be avoided in childhood and adolescence, as there is a tendency to compare them with normal young adult data, to misinterpret the Z scores, to report osteoporosis when none exists, and to overlook discrepancies in bone age. If needed, a bone mineral density in a child or adolescent should be done in a pediatric center familiar with these considerations. Fractures do occur at this age, and reflect rate of bone growth and vulnerable skeletal integrity. Some fractures occur simply due to excessive force or trauma. Studies have been published that address the question of whether childhood fracture predicts a fragility fracture later in life [17, 18]. The answer seems to be no.

Adolescence is a complicated time for all children but marked by significant illness in some. Depression, neuromuscular diseases (cerebral palsy), cancer, asthma, and juvenile inflammatory arthritis occur at this age; obesity is epidemic. These may result in inactivity, compromised height (steroids or chemotherapy), and complications of growth plates and modeling (micrognathia, leg length discrepancies, slipped capital femoral epiphysis, scoliosis, and spondylolisthesis). Endocrinopathies (delayed onset menses), inherited disorders of connective tissue (Marfan's), and storage diseases (Gaucher's) may all complicate skeletal growth and development.

As we do not know how to duplicate this phase of bone growth and bone mass accrual that occurs in early adolescence, these diseases may have devastating events in bone health. The safety of bisphosphonates has a mixed record in this population, diminishing fractures in some (osteogenesis imperfecta) [19, 20] and improving bone mineral density in others (anorexia nervosa) [13, 14]. Whether this will translate into significant gains in bone quality or in bone health in adults remains unclear [21]. It should be stressed that despite efficacy by the measurement of some parameters, e.g., bone mineral density or decreased fracture risk, there is no study on long-term safety in this vulnerable population [22, 23].

The importance in thinking of bone as vital tissue that is constantly remodeled is exemplified by this use of bisphosphonates in young women. These drugs have a long half-life in bone and are gradually released from the bone matrix over months to years. Their mechanism of action is usually one of inhibiting farnesyl synthase, an enzyme critical in the prenylation of small G proteins. This interferes with cyto-skeletal rearrangement necessary for bone resorption. As we learn about bone turn-over, we are beginning to understand that osteoclasts play an important role in balancing the skeletal needs for repair, the body's needs for calcium, and the signals involved in bone formation [24]. Disturbing this interplay of bone cells with their environment should be done cautiously in the young and with specific endpoints of easing disease manifestations such as fragility fracture, rather than surrogate mark-ers of bone turnover such as bone mineral density or bone turnover markers.

The creep of bisphosphonates into the premenopausal population where they are used for non-life-threatening diseases – such as accelerated bone loss in inflamma-tory rheumatic diseases and in patients with glucocorticoid-induced bone loss – has been largely unstudied in these terms. The bisphosphonates are classified as cate-gory C drugs by the FDA (alendronate (Fosamax), risedronate (Actonel), zoledronic acid 4 mg (Zometa)) and category D (pamidronate (Aredia), zoledronic acid 5 mg (Reclast)). As the FDA writes "There are no adequate and well-controlled studies in pregnant women." Animal studies do show these drugs pass through the placenta and are adsorbed by fetal skeletons. Although the only data we have on pregnancy in humans are scattered case reports, there remains a serious risk of harm intuited from animal data. Women should be counseled not to become pregnant while taking these drugs [25–28]. There is a brief discussion of this by Dr. Susan Ott in the http://courses.washington.edu/bonephys web pages under Normal Pregnancy in which these case studies are cited.

Pregnancy itself does not confer a risk for lifelong osteoporosis in most women. Calcium recruited from the maternal skeleton for fetal and then newborn health seems rapidly restored as breast-feeding ends. Rarely, regional or systemic osteopo-rosis complicates pregnancy in a woman.

Premenopause

It is clear that bone loss is occurring in some women before menopause. Premenopausal fractures may indicate risk or predict subsequent fracture in this still young population [29, 30]. Estrogen deficiency might be treated with hormone replacement therapy until the normal age of menopause (about 52 years) if there is no contraindication. Drugs enhancing bone loss should be identified – aromatase inhibitors, proton pump inhibitors, steroids, and thiazolidiones – and discontinued when possible. Calcium must be sufficient in the diet and supplemented when not. Calcium is best taken with food. Malabsorption should be sought as one factor that

Table 3.1 Some determinants of bone loss and fracture risk

Age
Caucasian
Maternal hip fracture
Genetics
Endocrinopathies
Hyperparathyroidism
Hyperthyroidism
Cushing's disease
Rheumatoid arthritis
Anemias, thalassemias
Inflammatory bowel disease
Drugs
Glucocorticoid are steroids
Cancer therapies
Proton pump inhibitors
Cigarettes
Alcohol
Weight loss >10 lbs
Vitamin D deficiency
Malabsorption
Celiac sprue
Risk of falls
Initiation of blood pressure medications
Institution of narcotics
Stroke
Instability of gait
Muscle weakness

may contribute to impaired nutrition and mineralization. Exercise, reasonable sunlight, and vitamin D 800–1,000 IU daily should be ensured. A dogmatic approach to low bone mineral density reports should be avoided. Fracture is less common in younger women, and evaluation for secondary causes of osteoporosis should be aggressive (Table 3.1).

Postmenopause

The issue that brought bone health to the forefront of medicine was the emergence of drugs effective in the treatment of osteoporosis. According to the World Health Organization, there are an estimated 8.9 million fractures annually in the world. As America ages, this risk of fracture and the attendant morbidity and mortality it incurs are the subject of the rest of this chapter. The premise in this discussion is an understanding that postmenopausal bone is not the same as young bone and that in menopause with the loss of estrogen, there is enhanced remodeling with imperfect bone formation in response to bone resorption. Adipocytes become more prevalent

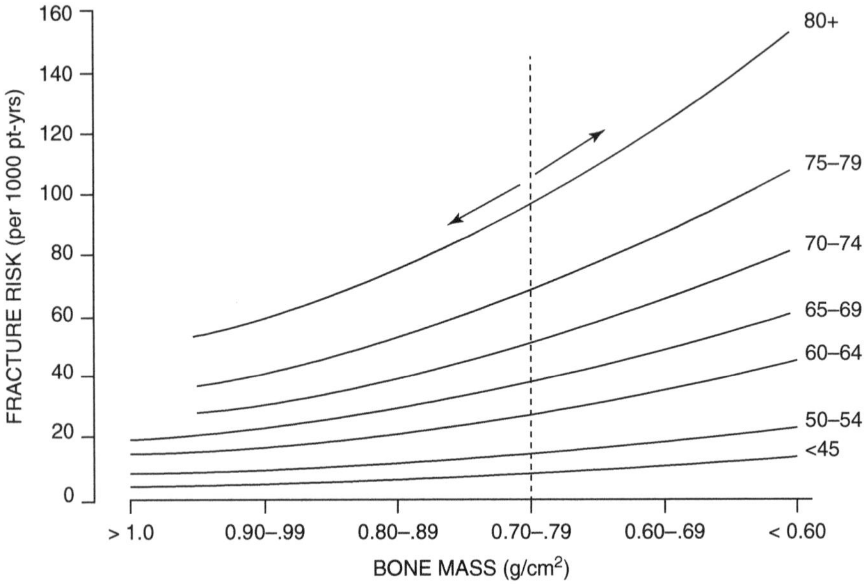

Fig. 3.1 Age as a determinant of fracture risk plotted against bone mineral density (Used with permission from Heaney [76])

in bone marrow and seem to play a role in driving some aspects of bone loss with aging [31]. Genetics play a significant role in the development of osteoporosis, and the concomitants of aging – osteoarthritis, gait instability, muscle weakness, and medications resulting in altered mental status or blood pressure changes – contribute to falls. In this population, intervention is safe and effective and should be undertaken.

Osteoporosis is defined as a skeletal disorder in which impaired microarchitecture of bone leads to a higher risk of fracture. Perhaps a natural process of aging, this has become a significant health problem as women live longer. Nearly 40% of Caucasian women will suffer an osteoporotic fracture in her lifetime; The ability to diagnose osteoporosis by bone mineral density has given physicians a chance to alter the consequences of this by improving bone strength with medication, diminishing bone loss, and thus decreasing the risk of fracture. Bone mineral density has proven an excellent correlate for fracture when interpreted in light of the patient's age and risk factors for fracture (Fig. 3.1) [32, 33].

Bone mineral density (BMD) interpretation depends on accurate positioning and appropriate interval testing to ensure a meaningful interpretation of the test. This is because bone loss is subtle in osteoporosis, and intervals of testing <2 years in untreated women will tend to result in variation rather than true difference. Interpreting bone mineral density changes from readings done on different BMD machines is difficult. A recent study on the indication for repeat testing by BMD argued that for those women with early osteopenia or normal BMD, the study might be repeated every 15 years rather than every 2 years [34]. This might be a bit

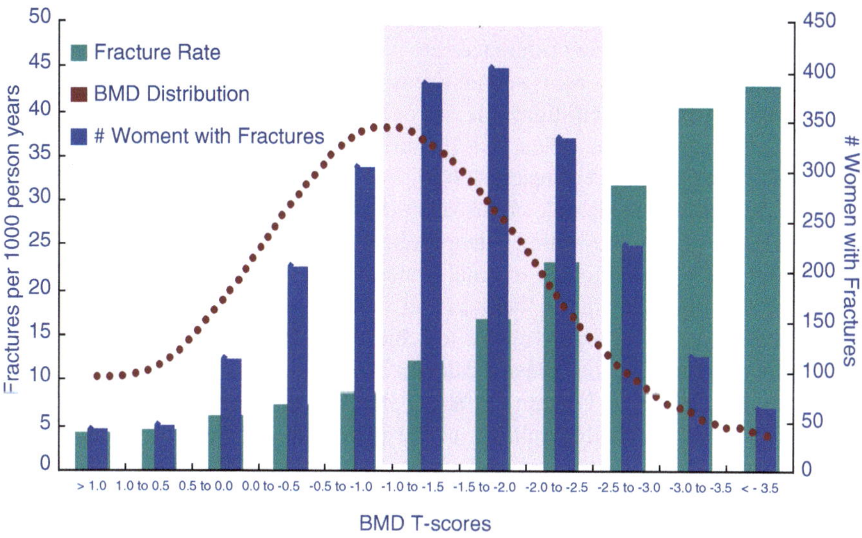

Fig. 3.2 Fracture rates in women and women with fracture as plotted against BMD (Used with permission from Siris et al. [35])

dogmatic, as some women may be edging towards an indication for treatment based upon fracture risk assessment (see below), and it may be prudent to repeat BMD more frequently. While rates of fracture increase with age, the number of women with extant fractures are those with osteopenia. A good discussion on the evaluation and treatment of osteopenia may be found in a clinical case study written to address this in the *New England Journal of Medicine* 2007 (Fig. 3.2) [35, 36].

Wrist fractures occur early in postmenopausal women, followed by rising rates of spinal compression fractures by age 65 and hip fractures by 70. The epidemiology of fracture accounts for the recommendations by the National Osteoporosis Foundation that a woman be screened for osteoporosis at age 65 by BMD, earlier if she has identifiable risk factors for fracture. Prior fracture, increasing age and maternal history of hip fracture are some of these indications for early screening. In an effort to calculate the absolute risk of hip and other fractures in women, the FRAX was developed, which is an online fracture risk assessment tool that uses the BMD at the hip, the patient's age, and corresponding risk factors to develop a 10-year risk of fracture [77]. While FRAX is an imperfect tool, since it fails to assess bone loss in the spine and fails to account for prior treatment, it is useful for creating a threshold for treatment. FRAX also helps women put into real terms their own personal fracture risk.

It is easy to treat a postmenopausal woman for osteoporosis, and yet our inclination to do so seems marginal [37, 38]. Even if we do write out a prescription for a bisphosphonate, compliance with the medication is poor without good communication about goals and length of treatment with the patient. In the USA, bisphosphonates, raloxifene, estrogen, denosumab, and teriparatide are available for the treatment of osteoporosis. Their indications, complications, and controversies are reviewed in Table 3.2, where references are provided for each of these issues.

Vitamin D and calcium are essential to ensure the proper mineralization of bone, but by themselves will not mitigate the risk of bone loss in the early postmenopausal years. There may be some efficacy in fracture reduction in their use in the elderly [67]. This is an active area of research [68].

There has been considerable controversy in the literature about the role vitamin D and calcium play in preventing osteoporotic fractures and the risk of supplemental calcium in the epidemiology of renal stones and cardiovascular disease [69–72]. Despite these arguments in the literature, all women deserve adequate nutrition for bone mineralization. If they are unable to achieve this through dietary calcium, then supplements are needed to achieve a calcium balance near 1,200 mg daily for older adult women. Generally, vitamin D3 800–1,000 IU daily reflects adequate intake. This information is readily available to the public on the National Osteoporosis Foundation website (www.nof.org) and the National Institutes of Health, Osteoporosis and Related Bone Diseases (www.niams.nih.gov).

Many patients with end-stage renal disease have poor bone quality, and some have adynamic bone disease. Antiresorptive drugs – bisphosphonates and denosumab – are probably poor choices in the setting of low bone turnover. Concurrent metabolic problems such as hyperparathyroidism and elevated FGF23 indicate other insults to bone. Teriparatide is not an option in this setting. While renal transplant patients are routinely given intravenous bisphosphonates prior to transplantation in an effort to mitigate the high risk of posttransplant fracture, there is no clear physiologic basis for this [73, 74].

In postmenopausal women, glucocorticoid therapy is associated with a high risk of fracture, particularly vertebral compression fracture, even in those with normal bone mineral density. Bisphosphonates and teriparatide [66] have demonstrated efficacy in mitigating this fracture risk and should be instituted within weeks to months of initiating therapy. Controversy exists around the wisdom of using antiresorptive agents such as the bisphosphonates in such settings [75], but these theoretical concerns have not yet been translated into clinical practice guidelines or current recommendations.

As "women hold up half the sky," they have been a vulnerable target for drug companies. There is direct marketing to the consumer. Estrogen, once on the cover of Time magazine as "every woman's dilemma" June 1995, was transformed from a positive intervention to one filled with more risk than benefit. In addition, the rapid newspaper coverage of toxicities and the changing impressions of the academic community on the wisdom of estrogen, calcium, and vitamin D use have left both patients and practitioners bewildered. Understandably, women have become more skeptical of these medications due to post-marketing reports of complications (atypical femoral fractures, osteonecrosis of jaw). Still, in the postmenopausal population, there is much good to be gained from prudent use of these drugs.

By thinking about bone as a living tissue and by understanding the impact of drugs on bone and the controversies that surround bone health, it is possible to weigh the benefits and risks of each of these medications in a woman's life. My own

Table 3.2 Overview of medications in postmenopausal women with osteoporosis

Bisphosphonates	Alendronate, risedronate, and zoledronic acid
Indications	Postmenopausal osteoporosis, prevention of osteoporosis, glucocorticoid-induced osteoporosis, Paget's disease of bone
	Ibandronate is indicated in the treatment and prevention of osteoporosis, proving most efficacious in reducing vertebral fractures; unproven in reducing hip fractures [39]
What recommends these?	Inexpensive, easy oral regimens and effective in osteoporosis
	Generally expect a 60 % reduction in vertebral fractures, 40 % reduction in hip fractures; and improvement in BMD
Randomized clinical trials	Yes, demonstrating efficacy and safety [40–42]
Complications	Common: flu-like symptoms with initial use, more common with infusions; bone pain; asymptomatic, transient lowering of serum calcium; GI distress/esophageal ulcers with oral medications; unstudied in patients with CrCl <30 ml/min
	Uncommon: osteonecrosis of the jaw
	Rare: atypical femoral fracture [43, 44]
	Acute renal failure [45]
	Iritis [46]
Controversy	Atrial fibrillation, both negative and positive findings [47, 48]
	Esophageal cancer: rare cancer in women; data conflicting, no proof to date [49, 50]
Ease of administration	Poor oral absorption: need to take on rising with 8 Oz of water and remain upright with no food or drink for 30–60 min
Selective estrogen receptor modulator	Raloxifene
Indications	Osteoporosis, treatment and prevention
	Postmenopausal breast cancer, treatment and prevention
What recommends this?	Inexpensive, well-tolerated, breast cancer prevention trials, cardiac safety
	Reduces vertebral fractures, no proven efficacy in hip fractures
Randomized clinical trials	Yes, demonstrating efficacy and safety [51, 52]
Complications	Thromboembolic disease, hot flashes, vaginitis, stroke in women with known coronary artery disease
Calcitonin	
Indication	Postmenopausal osteoporosis
Randomized clinical trials	Yes [53, 54]
Complications	Nausea, flushing, hypotension, allergic reaction, nasal irritation
Controversy	Increased prevalence of malignancy [55]
Ease of administration	Nasal spray generally well tolerated
Hormone replacement therapy	Estrogen with or without progesterone [56, 57]
Indications	Postmenopausal symptoms (indicated), prevention of postmenopausal osteoporosis (no longer advised as sole reason for use)
What recommends this?	Inexpensive, effective, eases postmenopausal symptoms
Randomized clinical trials	Yes
Complications	Thromboembolic disease, [58] cardiovascular disease, [59] invasive breast cancer [60]
	No protection against Alzheimer's or colon cancer
	Effective in reducing risks of hip and vertebral fractures
Rank-ligand antibody	Denosumab

(continued)

Table 3.2 (continued)

Indications	Postmenopausal osteoporosis in women with high risk of fracture, prevention of bone metastases in patients with solid tumors; giant cell tumor; and settings in which androgen/estrogen deprivation therapy is used for cancer therapy
What recommends this?	Effective in postmenopausal osteoporosis, metabolism safe in renal disease (although not clear it is beneficial in this setting)
Randomized clinical trials	Yes [61]
Complications	Eczema, cellulitis, osteonecrosis of jaw [62]
	Uncommon: profound hypocalcemia [63]
	6-month duration, then rebound of bone resorption
	Expensive
Ease of administration	By injection every 6 months
Parathyroid	Teriparatide
Indications	Osteoporosis (severe), glucocorticoid-induced osteoporosis
What recommends this?	Anabolic agent, effective in building bone and reducing fracture
	Generally expect a 60 % reduction in vertebral fractures and 40 % reduction in hip fractures and improvement in BMD
	Effect transient, may be more potent in combination with other drugs used for osteoporosis
	Safety in 7-year surveillance good [64]
Randomized clinical trials	Yes [65, 66]
Complications	Contraindications: prior radiation therapy; Paget's disease of bone, cancer metastatic to bone; hyperparathyroidism; others
	Common: transient lowering of blood pressure, elevations in serum calcium, nausea
	Rare: osteosarcoma (in rat models)
	Expensive
Ease of administration	Daily by injection for 2 years

recommendation is to limit skeletal exposure to these long-acting bisphosphonates, consider drug holidays after 3–5 years in the postmenopausal woman with osteoporosis, and reassess the need for these drugs at all in women of childbearing years. The references are meant as an updated guide to these issues.

References

1. Rauch F. Fetal and neonatal bone development. In: Rosen CJ, editor. Primer in metabolic bone disease. Washington, D.C: American Society for Bone and Mineral Research; 2009. p. 72–4.
2. Pludowski P, Holick MF, Pilz S, Wagner CL, Hollis BW, Grant WB, et al. Vitamin D effects on musculoskeletal health, immunity, autoimmunity, cardiovascular disease, cancer, fertility, pregnancy, dementia and mortality-A review of recent evidence. Autoimmun Rev. 2013;12(10): 976–89. doi:10.1016/j.autrev.2013.02.004. pii: S1568-9972(13)00040-2.
3. Hollis BW, Wagner CL. Vitamin D and pregnancy: skeletal effects, nonskeletal effects, and birth outcomes. Calcif Tissue Int. 2013;92(2):128–39.
4. Dawodu A, Saadi HF, Bekdache G, Javed Y, Altaye M, Hollis BW. Randomized Controlled Trial (RCT) of vitamin D supplementation in pregnancy in a population with endemic vitamin D deficiency. J Clin Endocrinol Metab. 2013;98(6):2337–46.

5. Tobias JH, Cooper C. PTH/PTHrP activity and the programming of skeletal development in utero. J Bone Miner Res. 2004;19(2):177–82.
6. Javaid MK, Crozier SR, Harvey NC, Gale CR, Dennison EM, Boucher BJ, et al. Maternal vitamin D status during pregnancy and childhood bone mass at age 9 years: a longitudinal study. Lancet. 2006;367(9504):36–43 [Research Support, Non-U.S. Gov't].
7. Lawlor DA, Wills AK, Fraser A, Sayers A, Fraser WD, Tobias JH. Association of maternal vitamin D status during pregnancy with bone-mineral content in offspring: a prospective cohort study. Lancet. 2013;381(9884):2176–83.
8. Harvey NC, Javaid K, Bishop N, Kennedy S, Papageorghiou AT, Fraser R, et al. MAVIDOS Maternal Vitamin D Osteoporosis Study: study protocol for a randomized controlled trial. The MAVIDOS Study Group. Trials. 2012;13:13 [Multicenter Study Randomized Controlled Trial Research Support, Non-U.S. Gov't].
9. Solomon A. Far from the tree. New York: Scribner; 2012.
10. Chumlea WC, Schubert CM, Roche AF, Kulin HE, Lee PA, Himes JH, et al. Age at menarche and racial comparisons in US girls. Pediatrics. 2003;111(1):110–3 [Comparative Study Research Support, Non-U.S. Gov't Research Support, U.S. Gov't, P.H.S.].
11. Epelboym Y, Gendron RN, Mayer J, Fusco J, Nasser P, Gross G, et al. The interindividual variation in femoral neck width is associated with the acquisition of predictable sets of morphological and tissue-quality traits and differential bone loss patterns. J Bone Miner Res. 2012;27(7):1501–10 [Research Support, Non-U.S. Gov't Research Support, U.S. Gov't, Non-P.H.S.].
12. Schlecht SH, Jepsen KJ. Functional integration of skeletal traits: an intraskeletal assessment of bone size, mineralization, and volume covariance. Bone. 2013;56(1):127–38.
13. Misra M, Klibanski A. Bone health in anorexia nervosa. Curr Opin Endocrinol Diabetes Obes. 2011;18(6):376–82 [Research Support, N.I.H., Extramural Review].
14. Miller KK, Meenaghan E, Lawson EA, Misra M, Gleysteen S, Schoenfeld D, et al. Effects of risedronate and low-dose transdermal testosterone on bone mineral density in women with anorexia nervosa: a randomized, placebo-controlled study. J Clin Endocrinol Metab. 2011;96(7):2081–8 [Randomized Controlled Trial Research Support, N.I.H., Extramural].
15. Thein-Nissenbaum J. Long term consequences of the female athlete triad. Maturitas. 2013; 75(2):107–12.
16. Walsh JS, Eastell R, Peel NF. Effects of Depot medroxyprogesterone acetate on bone density and bone metabolism before and after peak bone mass: a case-control study. J Clin Endocrinol Metab. 2008;93(4):1317–23 [Research Support, Non-U.S. Gov't].
17. Buttazzoni C, Rosengren B, Tveit M, Lanidn L, Nilsson J, Karlsson M. Does a childhood fracture predict low bone mass in young adulthood? A 27-year prospective controlled study. J Bone Miner Res. 2013;28(2):351–9.
18. Pye SR, Tobias J, Silman AJ, Reeve J, O'Neill TW. Childhood fractures do not predict future fractures: results from the European Prospective Osteoporosis Study. J Bone Miner Res. 2009;24(7):1314–8 [Multicenter Study Research Support, Non-U.S. Gov't].
19. Alcausin MB, Briody J, Pacey V, Ault J, McQuade M, Bridge C, et al. Intravenous pamidronate treatment in children with moderate-to-severe osteogenesis imperfecta started under three years of age. Horm Res Paediatr. 2013;79(6):333–40.
20. Land C, Rauch F, Travers R, Glorieux FH. Osteogenesis imperfecta type VI in childhood and adolescence: effects of cyclical intravenous pamidronate treatment. Bone. 2007;40(3):638–44 [Clinical Trial Research Support, Non-U.S. Gov't].
21. Meier RP, Ing Lorenzini K, Uebelhart B, Stern R, Peter RE, Rizzoli R. Atypical femoral fracture following bisphosphonate treatment in a woman with osteogenesis imperfecta–a case report. Acta Orthop. 2012;83(5):548–50 [Case Reports Research Support, Non-U.S. Gov't].
22. Whyte MP, Wenkert D, Clements KL, McAlister WH, Mumm S. Bisphosphonate-induced osteopetrosis. N Engl J Med. 2003;349(5):457–63 [Case Reports Research Support, Non-U.S. Gov't].
23. Rauch F, Travers R, Munns C, Glorieux FH. Sclerotic metaphyseal lines in a child treated with pamidronate: histomorphometric analysis. J Bone Miner Res. 2004;19(7):1191–3 [Case Reports Research Support, Non-U.S. Gov't].

24. Teti A. Bone development: overview of bone cells and signaling. Curr Osteoporos Rep. 2011; 9(4):264–73 [Review].
25. Stathopoulos IP, Liakou CG, Katsalira A, Trovas G, Lyritis GG, Papaioannou NA, et al. The use of bisphosphonates in women prior to or during pregnancy and lactation. Hormones (Athens). 2011;10(4):280–91 [Evaluation Studies Review].
26. Munns CF, Rauch F, Ward L, Glorieux FH. Maternal and fetal outcome after long-term pamidronate treatment before conception: a report of two cases. J Bone Miner Res. 2004;19(10):1742–5 [Case Reports Research Support, Non-U.S. Gov't].
27. McKenzie AF, Budd RS, Yang C, Shapiro B, Hicks RJ. Technetium-99m-methylene diphosphonate uptake in the fetal skeleton at 30 weeks gestation. J Nucl Med. 1994;35(8):1338–41 [Case Reports].
28. Patlas N, Golomb G, Yaffe P, Pinto T, Breuer E, Ornoy A. Transplacental effects of bisphosphonates on fetal skeletal ossification and mineralization in rats. Teratology. 1999;60(2):68–73 [Research Support, Non-U.S. Gov't].
29. Rozental TD, Deschamps LN, Taylor A, Earp B, Zurakowski D, Day CS, et al. Premenopausal women with a distal radial fracture have deteriorated trabecular bone density and morphology compared with controls without a fracture. J Bone Joint Surg Am. 2013;95(7):633–42 [Comparative Study].
30. Chevalley T, Bonjour JP, van Rietbergen B, Ferrari S, Rizzoli R. Fracture history of healthy premenopausal women is associated with a reduction of cortical microstructural components at the distal radius. Bone. 2013;55(2):377–83.
31. Nelson-Dooley C, Della-Fera MA, Hamrick M, Baile CA. Novel treatments for obesity and osteoporosis: targeting apoptotic pathways in adipocytes. Curr Med Chem. 2005;12(19): 2215–25 [Research Support, Non-U.S. Gov't Review].
32. Cooper C. The crippling consequences of fractures and their impact on quality of life. Am J Med 1997;103:12S–7S; discussion 7S–9S.
33. Cooper C, Melton 3rd LJ. Epidemiology of osteoporosis. Trends Endocrinol Metab. 1992;3(6):224–9.
34. Gourlay ML, Fine JP, Preisser JS, May RC, Li C, Lui LY, et al. Bone-density testing interval and transition to osteoporosis in older women. N Engl J Med. 2012;366(3):225–33 [Research Support, N.I.H., Extramural].
35. Siris ES, Chen YT, Abbott TA, Barrett-Connor E, Miller PD, Wehren LE, et al. Bone mineral density thresholds for pharmacological intervention to prevent fractures. Arch Intern Med. 2004;164(10):1108–12 [Research Support, Non-U.S. Gov't].
36. Khosla S, Melton 3rd LJ. Clinical practice. Osteopenia. N Engl J Med. 2007;356(22): 2293–300 [Research Support, N.I.H., Extramural Review].
37. Izuora KE, Alazraki N, Byrd-Sellers J, Tangpricha V, Nanes MS. Fracture assessment tool risk scores in bone density reports do not change physician prescribing behavior for osteoporosis. Am J Med Sci. 2011;342(1):5–8.
38. Cadarette SM, Katz JN, Brookhart MA, Levin R, Stedman MR, Choudhry NK, et al. Trends in drug prescribing for osteoporosis after hip fracture, 1995–2004. J Rheumatol. 2008;35(2): 319–26 [Research Support, N.I.H., Extramural Research Support, Non-U.S. Gov't].
39. Miller PD, McClung MR, Macovei L, Stakkestad JA, Luckey M, Bonvoisin B, et al. Monthly oral ibandronate therapy in postmenopausal osteoporosis: 1-year results from the MOBILE study. J Bone Miner Res. 2005;20(8):1315–22 [Clinical Trial Clinical Trial, Phase III Randomized Controlled Trial Research Support, Non-U.S. Gov't].
40. Black DM, Cummings SR, Karpf DB, Cauley JA, Thompson DE, Nevitt MC, et al. Randomised trial of effect of alendronate on risk of fracture in women with existing vertebral fractures. Fracture Intervention Trial Research Group. Lancet. 1996;348(9041):1535–41 [Clinical Trial Randomized Controlled Trial Research Support, Non-U.S. Gov't].
41. Harris ST, Watts NB, Genant HK, McKeever CD, Hangartner T, Keller M, et al. Effects of risedronate treatment on vertebral and nonvertebral fractures in women with postmenopausal osteoporosis: a randomized controlled trial. Vertebral Efficacy with Risedronate Therapy (VERT) Study Group. JAMA. 1999;282(14):1344–52 [Clinical Trial Multicenter Study Randomized Controlled Trial Research Support, Non-U.S. Gov't].

42. Black DM, Delmas PD, Eastell R, Reid IR, Boonen S, Cauley JA, et al. Once-yearly zoledronic acid for treatment of postmenopausal osteoporosis. N Engl J Med. 2007;356(18): 1809–22 [Multicenter Study Randomized Controlled Trial Research Support, N.I.H., Extramural Research Support, Non-U.S. Gov't].
43. Gedmintas L, Solomon DH, Kim SC. Bisphosphonates and risk of subtrochanteric, femoral shaft, and atypical femur fracture: a systematic review and meta-analysis. J Bone Miner Res. 2013;28(8):1729–37. doi:10.1002/jbmr.1893.
44. Meier RP, Perneger TV, Stern R, Rizzoli R, Peter RE. Increasing occurrence of atypical femoral fractures associated with bisphosphonate use. Arch Intern Med. 2012;172(12):930–6.
45. Chang JT, Green L, Beitz J. Renal failure with the use of zoledronic acid. N Engl J Med. 2003;349(17):1676–9; discussion –9. [Letter].
46. Siris ES. Bisphosphonates and iritis. Lancet. 1993;341(8842):436–7 [Case Reports Letter].
47. Erichsen R, Christiansen CF, Froslev T, Jacobsen J, Sorensen HT. Intravenous bisphosphonate therapy and atrial fibrillation/flutter risk in cancer patients: a nationwide cohort study. Br J Cancer. 2011;105(7):881–3 [Comparative Study Research Support, Non-U.S. Gov't].
48. Bunch TJ, Anderson JL, May HT, Muhlestein JB, Horne BD, Crandall BG, et al. Relation of bisphosphonate therapies and risk of developing atrial fibrillation. Am J Cardiol. 2009; 103(6):824–8.
49. Ho YF, Lin JT, Wu CY. Oral bisphosphonates and risk of esophageal cancer: a dose-intensity analysis in a nationwide population. Cancer Epidemiol Biomarkers Prev. 2012;21(6):993–5.
50. Dixon WG, Solomon DH. Bisphosphonates and esophageal cancer–a pathway through the confusion. Nat Rev Rheumatol. 2011;7(6):369–72 [Research Support, N.I.H., Extramural Research Support, Non-U.S. Gov't].
51. Ettinger B, Black DM, Mitlak BH, Knickerbocker RK, Nickelsen T, Genant HK, et al. Reduction of vertebral fracture risk in postmenopausal women with osteoporosis treated with raloxifene: results from a 3-year randomized clinical trial. Multiple Outcomes of Raloxifene Evaluation (MORE) Investigators. JAMA. 1999;282(7):637–45 [Clinical Trial Multicenter Study Randomized Controlled Trial Research Support, Non-U.S. Gov't].
52. Barrett-Connor E, Mosca L, Collins P, Geiger MJ, Grady D, Kornitzer M, et al. Effects of raloxifene on cardiovascular events and breast cancer in postmenopausal women. N Engl J Med. 2006;355(2):125–37 [Multicenter Study Randomized Controlled Trial Research Support, Non-U.S. Gov't].
53. Chesnut 3rd CH, Silverman S, Andriano K, Genant H, Gimona A, Harris S, et al. A randomized trial of nasal spray salmon calcitonin in postmenopausal women with established osteoporosis: the prevent recurrence of osteoporotic fractures study. PROOF Study Group. Am J Med. 2000;109(4):267–76 [Clinical Trial Comparative Study Multicenter Study Randomized Controlled Trial].
54. Binkley N, Bolognese M, Sidorowicz-Bialynicka A, Vally T, Trout R, Miller C, et al. A phase 3 trial of the efficacy and safety of oral recombinant calcitonin: the Oral Calcitonin in Postmenopausal Osteoporosis (ORACAL) trial. J Bone Miner Res. 2012;27(8):1821–9 [Clinical Trial, Phase III Randomized Controlled Trial Research Support, Non-U.S. Gov't].
55. In brief: cancer risk with salmon calcitonin. Med Lett Drugs Ther. 2013;55(1414):29 [News]
56. Lobo R. Where are we 10 years after the women's health initiative? J Clin Endocrinol Metab. 2013;98:1771–80.
57. Utian WH, Archer DF, Bachmann GA, Gallagher C, Grodstein F, Heiman JR, et al. Estrogen and progestogen use in postmenopausal women: July 2008 position statement of The North American Menopause Society. Menopause. 2008;15(4 Pt 1):584–602 [Practice Guideline].
58. Archer DF, Oger E. Estrogen and progestogen effect on venous thromboembolism in menopausal women. Climacteric. 2012;15(3):235–40 [Research Support, Non-U.S. Gov't Review].
59. Miller VM, Manson JE. Women's health initiative hormone therapy trials: new insights on cardiovascular disease from additional years of follow up. Curr Cardiovasc Risk Rep. 2013;7(3):196–202.
60. Chlebowski RT, Manson JE, Anderson GL, Cauley JA, Aragaki AK, Stefanick ML, et al. Estrogen plus progestin and breast cancer incidence and mortality in the Women's Health Initiative Observational Study. J Natl Cancer Inst. 2013;105(8):526–35 [Multicenter Study].

61. Cummings SR, San Martin J, McClung MR, Siris ES, Eastell R, Reid IR, et al. Denosumab for prevention of fractures in postmenopausal women with osteoporosis. N Engl J Med. 2009;361(8): 756–65 [Multicenter Study Randomized Controlled Trial Research Support, Non-U.S. Gov't].
62. Denosumab. Limited efficacy in fracture prevention, too many adverse effects. Prescrire Int. 2011;20(117):145–8 [Comparative Study]
63. Milat F, Goh S, Gani LU, Suriadi C, Gillespie MT, Fuller PJ, et al. Prolonged hypocalcemia following denosumab therapy in metastatic hormone refractory prostate cancer. Bone. 2013;55(2):305–8.
64. Andrews EB, Gilsenan AW, Midkiff K, Sherrill B, Wu Y, Mann BH, et al. The US postmarketing surveillance study of adult osteosarcoma and teriparatide: study design and findings from the first 7 years. J Bone Miner Res. 2012;27(12):2429–37.
65. Neer RM, Arnaud CD, Zanchetta JR, Prince R, Gaich GA, Reginster JY, et al. Effect of parathyroid hormone (1–34) on fractures and bone mineral density in postmenopausal women with osteoporosis. N Engl J Med. 2001;344(19):1434–41 [Clinical Trial Multicenter Study Randomized Controlled Trial Research Support, Non-U.S. Gov't].
66. Saag KG, Shane E, Boonen S, Marin F, Donley DW, Taylor KA, et al. Teriparatide or alendronate in glucocorticoid-induced osteoporosis. N Engl J Med. 2007;357(20):2028–39 [Comparative Study Multicenter Study Randomized Controlled Trial Research Support, Non-U.S. Gov't].
67. Chapuy MC, Pamphile R, Paris E, Kempf C, Schlichting M, Arnaud S, et al. Combined calcium and vitamin D3 supplementation in elderly women: confirmation of reversal of secondary hyperparathyroidism and hip fracture risk: the Decalyos II study. Osteoporos Int. 2002;13(3):257–64 [Clinical Trial Multicenter Study Randomized Controlled Trial Research Support, Non-U.S. Gov't].
68. Dawson-Hughes B. What is the optimal dietary intake of vitamin D for reducing fracture risk? Calcif Tissue Int. 2013;92(2):184–90 [Research Support, U.S. Gov't, Non-P.H.S.].
69. Moyer VA. Vitamin D, and calcium supplementation to prevent fractures in adults: U.S. preventive services task force recommendation statement. Ann Intern Med. 2013;158(9):691–6 [Research Support, U.S. Gov't, P.H.S.].
70. Reid IR, Bolland MJ. Does widespread calcium supplementation pose cardiovascular risk? Yes: the potential risk is a concern. Am Fam Physician. 2013;87(3):Online [Editorial]
71. Lutsey PL, Michos ED. Vitamin D, calcium, and atherosclerotic risk: evidence from serum levels and supplementation studies. Curr Atheroscler Rep. 2013;15(1):293 [Research Support, N.I.H., Extramural Review].
72. Haghighi A, Samimagham H, Gohardehi G. Calcium and vitamin d supplementation and risk of kidney stone formation in postmenopausal women. Iran J Kidney Dis. 2013;7(3):210–3.
73. Kasturi KS, Chennareddygari S, Mummadi RR. Effect of bisphosphonates on bone mineral density in liver transplant patients: a meta-analysis and systematic review of randomized controlled trials. Transpl Int. 2010;23(2):200–7 [Meta-Analysis].
74. Conley E, Muth B, Samaniego M, Lotfi M, Voss B, Armbrust M, et al. Bisphosphonates and bone fractures in long-term kidney transplant recipients. Transplantation. 2008;86(2):231–7 [Research Support, N.I.H., Extramural].
75. Teitelbaum SL, Seton MP, Saag KG. Should bisphosphonates be used for long-term treatment of glucocorticoid-induced osteoporosis? Arthritis Rheum. 2011;63(2):325–8.
76. Heaney RP. Nutrition and osteoporosis. In: Favus MJ, editor. Primer on the metabolic bone diseases and disorders of mineral metabolism. 3rd ed. Philadelphia: Lippincott-Raven; 1996.
77. McCloskey EV, Johansson H, Oden A, Kanis JA. From relative risk to absolute fracture risk calculation: the FRAX algorithm. Curr Osteoporos Rep 2009;7:77–83.

Chapter 4
Upper Limb Nerve Entrapment Syndromes

Nicholas K. Muraoka and Jennifer Baima

Abstract The median nerve is formed by contributions from the lateral and medial cords of the brachial plexus. Signs and symptoms of a median neuropathy include dysesthesias of the first four fingers in a median distribution, hand pain, and thumb weakness, or atrophy of the thenar eminence. Median nerve compression at the wrist, or carpal tunnel syndrome, is the most common nerve entrapment syndrome. Pregnancy and hormonal factors may contribute to median nerve compression in the female wrist.

The second most common nerve compression site is the ulnar nerve at the cubital tunnel. Anywhere from 3 to 20 mm after the ulnar groove, the nerve runs in the *cubital tunnel*, formed beneath the two heads of the flexor carpi ulnaris muscle. Signs and symptoms consistent with ulnar neuropathy include intrinsic hand muscle weakness resulting in loss of dexterity and grip strength. Sensory symptoms are not as prominent as motor symptoms in an ulnar neuropathy. If present, the area affected will be the dorsal fifth and medial fourth digits as well as the medial hand. Typically, males have a higher prevalence of ulnar neuropathy than females.

The posterior cord of the brachial plexus gives off the axillary, thoracodorsal, and subscapular nerves before terminating as the radial nerve. Radial neuropathy at the spiral groove will present as wrist- and finger-drop but spares elbow extension since the muscular branches to the triceps brachii and anconeus muscles arise proximally to the spiral groove. In a superficial radial sensory neuropathy, there will be

N.K. Muraoka, DO
Department of Physical Medicinal and Rehabilitation,
Spaulding Rehabilitation Hospital, Boston, MA 02114, USA
e-mail: nmuraoka@partners.org

J. Baima, MD (✉)
Division of Women's Health, Department of Medicine,
Harvard Medical School, Brigham and Women's Hospital,
Chestnut Hill, MA 02467, USA
e-mail: jbaima1@partners.org

E. Mody, E. Matzkin (eds.), *Musculoskeletal Health in Women*,
DOI 10.1007/978-1-4471-4712-1_4, © Springer-Verlag London 2014

sensory disturbance of the lateral dorsal hand and thumb plus the dorsal proximal phalanges of digits 2, 3, and 4.

The suprascapular nerve comes off the upper trunk of the brachial plexus proximally and contains C5 and C6 innervation. The nerve runs posteriorly under the trapezius and travels through the suprascapular notch of the scapula into the supraspinous fossa to innervate the supraspinatus muscle. Volleyball, tennis, dancing, painting, pitching, and other overhand throwing are activities that have been associated with suprascapular neuropathy. There is no known gender association.

Traumatic injuries to the brachial plexus are the most common etiology of brachial plexopathies. Breast cancer, lymphomas, and lung cancer are the most frequent cause of lymphadenopathy causing plexopathy. Radiation treatment protocols often include the region of the brachial plexus, especially for treatment of lymphomas and breast, lung, and neck cancers. Idiopathic brachial plexopathy is more common in males. Nonneoplastic mass effects may occur, such as hematomas from internal jugular catheters or vascular abnormalities like arteriovenous malformations or aneurysms.

Keywords Median neuropathy • Carpal tunnel syndrome • Ulnar neuropathy • Radial neuropathy

Introduction

Many nerves exit the cervical spine and cross in the brachial plexus to innervate the muscles and skin of the shoulder and arm. Nerves can malfunction from direct compression or a metabolic or toxic process. This chapter will focus on compressive syndromes of the upper limb. We will review the most common nerve entrapment problems, carpal tunnel syndrome (median nerve) and cubital tunnel syndrome (ulnar nerve), as well as the less common radial nerve entrapment. We will explain the more common presentations of brachial plexopathy and how these problems may be different in women than men.

Median Neuropathy

Anatomy

The median nerve is formed by contributions from the lateral and medial cords of the brachial plexus. Individually, the *lateral cord* is made up of C6–C7 fibers and provides sensory fibers to the thenar eminence, thumb, index, and middle fingers and motor fibers to the proximal median forearm muscles. The *medial cord* is made up of C8–T1 fibers and provides sensory fibers to lateral half of the ring finger and motor fibers to the median muscles of the distal forearm and hand.

The median nerve is located medial to the humerus in the arm and lies anterior to the medial epicondyle of the humerus. In the antecubital fossa, the median nerve runs with the brachial artery and dives deep to the *lacertus fibrosis*, which is a thick fibrous band that runs from the medial aspect of the biceps tendon to the proximal forearm flexors. The median provides innervation to the pronator teres as it travels between the two heads of the muscle at the medial epicondyle. The median nerve also innervates the flexor digitorum sublimis and, in some, the palmaris longus muscle. The median nerve continues distally in the forearm and gives motor control to the flexor carpi radialis and the flexor digitorum sublimis before it runs below the tendinous origin of the flexor digitorum sublimis (*sublimis bridge*).

The *ligament of Struthers* is an anatomical variant that may have some clinical significance. This is a fibrous band between the medial epicondyle and an accessory supracondylar bony prominence at the distal medial humeral shaft that exists in about 1–2 % of the population. Clinically, this can cause an entrapment syndrome of the median nerve as well as the ulnar nerve.

The *anterior interosseous nerve* is a motor nerve that is given off posteriorly from the median nerve approximately 5–8 cm distal to the medial epicondyle. The anterior interosseous nerve innervates the medial head of the flexor digitorum profundus (index and middle fingers), flexor pollicis longus, and the pronator quadratus. The anterior interosseous nerve does not carry cutaneous sensory fibers, but does carry deep sensory fibers to the wrist and the interosseous membrane of the forearm.

The *palmar cutaneous sensory branch of the median nerve* is given off just proximal to the wrist and supplies sensation to the thenar eminence. At the wrist, the median nerve travels in the *carpal tunnel*, along with nine tendons (four from the flexor digitorum profundus, four from the flexor digitorum sublimis, and one from the flexor pollicis longus). See Fig. 4.1 for schematic of the anatomy of the carpal tunnel. In the palm, the median nerve splits into the *sensory branch* that innervates the medial thumb, index finger, middle finger, and lateral half of the ring finger via digital sensory branches and the *motor branch* that innervates the first and second lumbricals before giving off the recurrent thenar motor branch, which innervates the opponens pollicis, the abductor pollicis brevis, and the superficial head of the flexor pollicis brevis.

Possible sites of entrapment of the median nerve include: (1) the medial and lateral aspect of digits 1–3 and lateral aspect of digit 4; (2) at the carpal tunnel, formed by the transverse carpal ligament passing over the carpal bones at the wrist; (3) at the ligament of Struthers, where it passes with the brachial artery into the forearm; (4) beneath the lacertus fibrosis at the antecubital fossa; (5) as it passes through the pronator teres in the proximal forearm; and (6) beneath the sublimis bridge in the proximal forearm. Clinically, *pronator syndrome* refers to median nerve entrapment at any of the last three sites (beneath lacertus fibrosis, in the pronator teres muscle, and beneath the sublimis bridge). Of note, median nerve entrapment at the wrist in the carpal tunnel is the most common of all entrapment neuropathies [1].

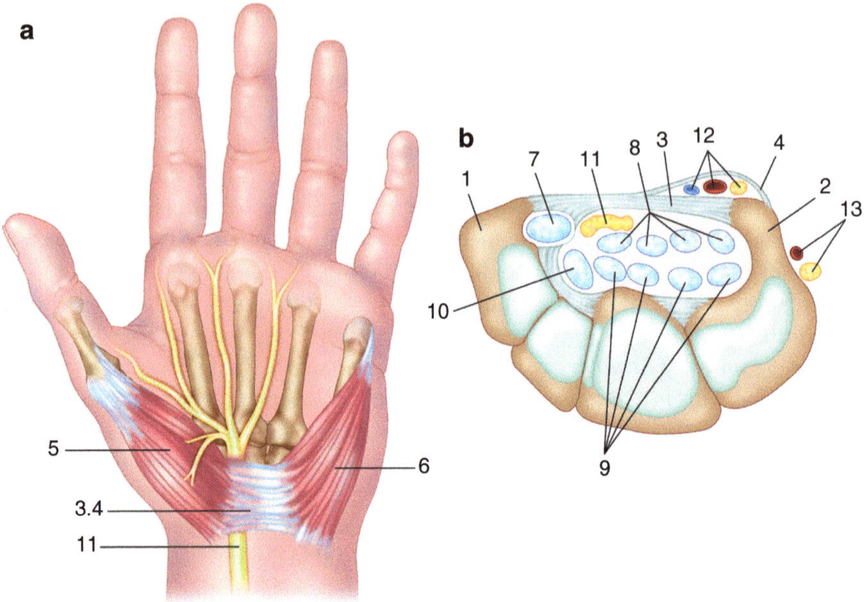

Fig. 4.1 (**a**, **b**) Anatomy of the carpal canal. (**a**) Typical median nerve and hand anatomy at the level of the transverse carpal ligament. (**b**) Carpal tunnel cross-sectional anatomy. *1* Trapezium tubercle, *2* hook of the hamate, *3* transverse carpal ligament, *4* palmar carpal ligament, *5* thenar muscles, *6* hypothenar muscles, *7* flexor carpi radialis tendon, *8* flexor digitorum superficialis tendon, *9* flexor digitorum profundus tendon, *10* flexor pollicis longus tendon, *11* median nerve, *12* ulnar artery, vein, and nerve superficial branches, *13* ulnar artery, vein, and nerve deep branches (Used with permission from Kang and Gupta [65])

Epidemiology

A population-based retrospective study in Rochester, MN, over 19 years identified the incidence of carpal tunnel syndrome (CTS) in females to be 149 per 100,000 versus only 52 per 100,000 in males, which puts the ratio of females to males at 3:2 [2]. A subsequent study of the same population revealed several conditions that were found with increased prevalence in patients with CTS compared to the general population. These conditions include Colles' fracture, rheumatoid arthritis, use of hormonal agents, hypothyroidism with myxedema, diabetes, and for males occupations that involved excessive use of the hands. Highest associations were rheumatoid arthritis (3.6 more likely), diabetes mellitus (2.3 more likely), and pregnancy (2.5 more likely).

However, despite these recognized associations, 43 % of the individuals with CTS did not have any of the conditions. This is termed idiopathic CTS [3]. Obesity is another risk factor for CTS, found to be 2.5 times more likely in patients with BMI >29 compared to those under 29 [4]. Interestingly, a study found that women who used hormone replacement therapy were more likely to undergo carpal tunnel release surgery than those women who did not take HRT (OR 1.8) [5]. Multiple studies have failed to prove that computer use is associated with development of CTS [6, 7].

History and Clinical Presentation

Signs and symptoms consistent with a median neuropathy:

1. Dysesthesias of the first four fingers of the hand in a median distribution, worse at night
2. Hand pain
3. Weakness of the thumb or atrophy of the thenar eminence

Classically, carpal tunnel syndrome (CTS) typically presents as wrist and arm pain associated with paresthesias in the hand, commonly at night. The pain may be in just the wrist or may radiate into the forearm, arm, or shoulder. Symptoms of carpal tunnel can be provoked with prolonged wrist flexion or extension. CTS without pain is rare. The dysesthesias classically involve the medial thumb, index finger, middle finger, and lateral ring finger. However, the clinician should be aware that symptoms may involve the ulnar distribution. In a survey of 100 patients with electrodiagnostically confirmed CTS, the most common pattern of dysesthesias involved all digits of the hand (48 % of cases). The classic distribution of symptoms was present in only 45 % of the group [8]. It is important to note on physical exam whether the sensation to the thenar eminence is impaired because the thenar eminence is innervated by the palmar cutaneous sensory branch, which is given off the median nerve proximal to the carpal tunnel.

Weakness in arm pronation (pronator quadratus distally and pronator teres proximally), wrist flexion (flexor carpi radialis), and distal thumb flexion (flexor pollicis longus) would indicate a more median nerve lesion, possibly at the elbow. Pronator syndrome or entrapment under the ligament of Struthers, which is a fibrous band between the medial epicondyle and an accessory supracondylar bony prominence at the distal medial humeral shaft that exists in about 1–2 % of the population, can produce pain or paresthesias in the volar forearm and thenar eminence in addition to the same areas affected by CTS.

Anterior interosseous neuropathy (AIN) typically presents as an inability to flex the distal phalanges of the thumb, index, and middle fingers plus weak forearm pronation. There is no sensory loss since the anterior interosseous nerve does not carry cutaneous innervation.

Compression of the digital sensory nerves along the sides of digits 1–4 may cause pain or paresthesias in the distal finger. Baseball and bowling are sports that have a tendency to cause digital neuropathies. Flute players may experience this on the lateral side of left index finger. Violinists and cellists may be symptomatic in the right thumb. Percussionists tend to experience this in the left middle finger. Vibration exposure can also cause digital neuropathy [9].

Other diagnoses in the differential with median neuropathy are cervical radiculopathy and brachial neuritis. With cervical radiculopathy, there is usually a history of neck pain. Acute herniated discs can produce a radiculopathy with pain with concurrent sensory and motor symptoms. Brachial plexitis, also known as neuralgic amyotrophy or Parsonage-Turner syndrome, presents as severe arm and shoulder pain that is followed after a period of days by numbness and weakness.

Exam

The Phalen's maneuver involves passive flexion of the wrists, which will produce paresthesias in CTS within 30–120 s. Tinel's sign is the presence of paresthesias in the median distribution of the fingers with percussion over the carpal ligament at the palmar surface of the wrist. Phalen's maneuver is a slightly better test than Tinel's sign with reported sensitivity and specificity of 68 and 73 % versus 50 and 77 %, respectively, in a systematic review [10].

In pronator syndrome, there are several maneuvers that can be done to determine what structure is compressing the median nerve in the proximal forearm. A positive exam would be pain or paresthesias in a median distribution [11]:

- Resisted pronation with the elbow in full extension → pronator teres
- Resisted flexion of the PIP joint of the middle finger → sublimis bridge
- Resisted elbow flexion with forearm supinated → lacertus fibrosis

A bony prominence may be palpated at the distal medial humerus, suggesting the presence of a ligament of Struthers variant. Repeated supination and elbow extension would produce maximum contact of the ligament with the median nerve.

Anterior interosseous neuropathy will affect flexion of the DIP joint of the index and middle finger and flexion of the IP joint of the thumb. When a patient tries an "OK" sign, there will be compensatory hyperextension at these joints to complete a circle. Weakness of the pronator quadratus can be tested by resisted forearm pronation with the elbow flexed. When the elbow is extended, the pronator teres is the primary pronator.

The biceps and brachioradialis reflexes are important to examine. These median neuropathies of the wrist and forearm would have intact biceps, brachioradialis, and triceps reflexes (C6 and C7 reflexes). Impaired reflexes are consistent with a more proximal lesion, such as the brachial plexus or cervical nerve root level.

Diagnosis

Imaging does not have a significant role in diagnosis. CTS is a clinical diagnosis, meaning that there should be associated symptoms and objective exam findings. Electrodiagnostic studies can help confirm the clinical diagnosis. There may be electrodiagnostic evidence of slowing across the carpal ligament in a patient without symptoms or signs consistent with CTS. This may be encountered in someone who presents with numbness and tingling in hands and lower extremities who on testing shows nerve slowing across common entrapment sites in the setting of an underlying polyneuropathy, such as diabetic or alcoholic neuropathy. In these patients, the underlying metabolic or toxic cause of neuropathy should be addressed.

Electrodiagnostic studies should include a comparison study of the median nerve with a different nerve in the hand, usually the ulnar nerve and less commonly the radial nerve. Addition of the comparison study increases the sensitivity from 75 % to as high as 95 %. Median-to-ulnar comparison studies include palm-to-wrist (median vs. ulnar) mixed nerve latency, wrist-to-digit 4 (median vs. ulnar) sensory latency, and second lumbrical (median) to first dorsal interossei (ulnar) distal motor latency.

Treatment

In general, treatment of carpal tunnel syndrome is targeted at managing the painful symptoms and preventing a loss of function. Activity modification can be a first intervention. There is evidence that use of an ergonomic keyboard is effective in reducing symptoms of CTS after 3 months of use compared to use of a standard keyboard [12]. Physical therapy can be prescribed to increase range of motion in the wrist and address strength impairments. However, there is no evidence that therapeutic exercises alone are effective in treating carpal tunnel syndrome [13].

Splinting can be used to place the wrist in a comfortable position to minimize pressure to the median nerve within the carpal tunnel. It has been found that placing the wrist in a neutral position at night results in improvements in pain and function at 4 weeks [14]. An RCT comparing full-time use of a wrist splint to nighttime use only did not find a significant difference in outcomes, suggesting that splinting at night is adequate for improvement [15].

Oral medications can be used to treat symptoms. Oral steroids have been shown to be effective in the short term to improve symptoms of CTS [16]. NSAIDs are commonly prescribed for their analgesic and anti-inflammatory effects. However, there is limited evidence that NSAIDs are more effective than placebo in treating CTS. A review found that the use of nonsteroidal anti-inflammatory medications, vitamin B6, and diuretics was no more effective than placebo in treating CTS [17].

Steroid injections into the carpal tunnel by an experienced clinician can result in significant improvement. A Cochrane review concluded that there was significant short-term clinical improvement after a steroid injection [18]. Risks of a steroid injection include bleeding, infection, and nerve injury from iatrogenic injection into the nerve.

Surgical treatment of CTS should be reserved for individuals with evidence of moderate to severe neuropathy on electrodiagnostic studies or who are experiencing a functional decline despite nonsurgical treatment. Release of the transverse carpal ligament can be done by a surgeon using either an open approach or an endoscopic approach. There is no substantial evidence that one surgical approach is more effective than the other [19].

In pronator syndrome and anterior interosseous syndrome, relative rest, splinting, and anti-inflammatories should be attempted. If there is no improvement in symptoms after 6 months, surgical exploration can be considered. Usually, this leads to recovery unless severe axonotmesis is present [20]. Median nerve sensory fiber involvement in the digits can be difficult to treat. Initially, splinting and anti-inflammatories can be used in the acute phase. Steroid injection and/or surgical release is often the treatment of choice [9].

Ulnar Neuropathy

Anatomy

The ulnar nerve is formed by contributions from C8 and T1 roots. The lower trunk gives off the medial cord, which gives contributions to the medial pectoral nerve, and medial brachial cutaneous and medial antebrachial cutaneous nerves before becoming the ulnar nerve distally. The medial brachial cutaneous and medial antebrachial cutaneous nerves supply cutaneous sensation to the medial aspect of the arm and forearm, respectively.

The ulnar nerve initially travels with the median nerve and brachial artery in a neurovascular bundle before it passes from the medial arm through the medial intermuscular septum into the posterior compartment of the arm and through the *arcade of Struthers*. The arcade of Struthers contains fascia, the medial head of the triceps, and the internal brachial ligament. At the elbow, the ulnar nerve passes into the *ulnar groove*, which is formed by the medial epicondyle (humerus) and the olecranon process (ulna). Anywhere from 3 to 20 mm after the ulnar groove, the nerve runs in the *cubital tunnel*, formed beneath the two heads of the flexor carpi ulnaris muscle. Innervation is given to the flexor carpi ulnaris and the medial division of the flexor digitorum profundus (fourth and fifth digits).

The nerve then travels down the medial forearm and approximately 5–8 cm proximal to the wrist and gives the *dorsal ulnar cutaneous sensory branch*, which supplies sensation to the volar medial fourth and fifth digits and palmar cutaneous sensory branch. At the medial wrist, the nerve passes into *Guyon's canal*, which is formed by the pisiform bone and hook of the hamate. Within Guyon's canal, the ulnar nerve divides into a superficial sensory branch and a deep motor branch. The deep motor branch gives motor supply to the hypothenar muscles (abductor digiti minimi, opponens digiti minimi, flexor digiti minimi, and palmaris brevis) before exiting Guyon's canal. After exiting Guyon's canal at the pisohamate hiatus, the deep motor branch supplies all three palmar and four dorsal interossei; the third and fourth lumbricals, which flex the third and fourth digits at the metacarpophalangeal joint and extend them at both proximal and distal interphalangeal joints; and two muscles in the thenar eminence: the adductor pollicis and the deep head of the flexor pollicis brevis.

There are five sites of potential entrapment of the ulnar nerve: (1) within the fibrous arcade of Struthers proximal to the elbow; (2) at the ulnar groove, where the nerve travels between the medial epicondyle and olecranon; (3) at the cubital tunnel, where the ulnar nerve enters at the humeroulnar aponeurotic arcade (HUA) and travels to exit the tunnel at the deep flexor pronator aponeurosis; (4) at the wrist at Guyon's canal; and (5) in the palm where the ulnar nerve exits Guyon's canal as the deep and superficial branches.

Epidemiology

Ulnar neuropathy at the elbow is the second most common entrapment neuropathy site after carpal tunnel syndrome [21]. The mean annual crude incidence based on a population study in Italy is 24.7 cases per 100,000 people annually. In that study population, men were affected more than women by a ratio of 2:1 [22]. Prolonged elbow flexion may be a risk factor. In a study of musicians, ulnar neuropathy at the elbow was studied. In the musicians that used bowed instruments (violin, viola, etc.), there was only left-sided involvement. Since the playing position involves left elbow flexion and forearm supination, this position may be a risk factor for ulnar neuropathy at the elbow [23].

Males develop perioperative ulnar neuropathies more frequently. Anatomical studies reveal larger coronoid process in males as well as higher fat content at the medial elbow in females, suggesting mechanical compression susceptibility plus less protection as a risk factor accounting for gender differences [24]. Ulnar neuropathy at the wrist is usually localized to Guyon's canal and can be caused by bicycle riding, push-ups, flute or violin playing, ganglion cysts, ulnar artery aneurysm/thrombosis, and lipomas [25].

History and Clinical Presentation

Signs and symptoms consistent with ulnar neuropathy:

1. Weakness of the intrinsic hand muscles resulting in loss of dexterity and grip strength. Patients may report difficulty with writing, buttoning clothes, and opening jars.
2. Sensory symptoms are generally not as prominent as motor symptoms in an ulnar neuropathy. If present, the area affected will be the dorsal fifth and medial fourth digits as well as the medial hand.

Ulnar neuropathy at the elbow is usually the result of chronic mechanical compression or stretch of the nerve. At the *ulnar groove* its superficial location makes it vulnerable to external forces. The term "hitting the funny bone" relates to striking the ulnar nerve at the ulnar groove and causing paresthesias into the fourth and fifth digits. Some individuals can have an ulnar nerve that subluxes out of the groove over the medial epicondyle when they fully flex their elbows, possibly making the nerve more vulnerable to trauma.

A history of elbow fracture may predispose an individual to early arthritic changes to the elbow that can compromise the ulnar nerve along its course. This is referred clinically as *tardy ulnar palsy*. After the ulnar groove, the nerve dives into the *cubital tunnel*, which is formed by the HUA. Some individuals have narrow space within the cubital tunnel, which can predispose them to compression of the ulnar nerve within the tunnel. Mechanically, flexion at the elbow does decrease the diameter of the cubital tunnel and can exacerbate symptoms [26]. *Cubital tunnel syndrome* is used commonly to refer to all presentations of ulnar neuropathy at the elbow, although, technically, the term applies only to nerve involvement under the HUA and not involvement at the ulnar groove.

Ulnar neuropathy at the wrist is less common than involvement at the elbow. Areas of entrapment can be either within Guyon's canal or in the palm after the ulnar nerve exits Guyon's canal as a superficial sensory branch and deep motor branch. Clinical presentation may vary slightly, depending on if one or both of the branches are affected. However, 75 % of the cases will involve the motor branch only and present as weakness of the ulnar-innervated hand muscles without sensory loss with variable hypothenar eminence involvement. A history of hand fracture, particularly the hook of the hamate, can increase the risk of arthritic changes in the wrist that can contribute to ulnar nerve compression. Cases of ulnar neuropathy at the wrist can be due to a ganglion cyst within Guyon's canal. Also trauma or repetitive impact to the wrist as in bikers and laborers who use power tools may cause dysfunction at the wrist.

The differential diagnoses for suspected ulnar neuropathy includes a C8 or T1 radiculopathy and a lower trunk or medial cord plexopathy. Cervical radiculopathies usually present with a history of neck pain, and symptoms will be exacerbated with extension of the cervical spine. For ulnar neuropathy at the wrist presenting as motor impairment in the absence of sensory involvement, the list of differential diagnoses includes motor neuron disease, so it is important to make sure this is an isolated complaint. Medial epicondylitis may also present with medial elbow pain that can localize to the area near the ulnar groove and cubital tunnel.

Exam

On palpation, there may be reproduction of paresthesias with percussion at the medial elbow, consistent with entrapment at the ulnar groove or in the cubital tunnel. On sensory exam, sensation should be intact in the forearm. If the medial forearm has abnormal sensation, this could indicate a lesion at or proximal to the medial antebrachial cutaneous nerve, which branches off the medial cord prior to it becoming the ulnar nerve. Therefore, altered sensation in the forearm suggests a lesion of the medial cord, lower trunk, or contributing nerve roots (C8 and T1). Sensory exam should include the dorsal medial hand. This area is served by the dorsal ulnar cutaneous sensory nerve, which branches off the ulnar nerve proximal to the wrist. Therefore, if this area has impaired sensation, the lesion would be proximal to the wrist.

Patients with an ulnar neuropathy will have difficulty making a fist with the fourth and fifth digits due to weakness of the fourth and fifth flexor digitorum profundus, which are ulnar-innervated, in comparison to the FDP of the second and third digits, which are median-innervated. Thumb abduction is spared (median and radially innervated).

Benediction posture is the classic hand presentation with advanced ulnar neuropathy. The fourth and fifth digits are in a claw position, with MCP joints hyperextended and both interphalangeal joints flexed secondary to lumbrical weakness. The digits, including the thumb, are slightly abducted from weakness of the interossei and adductor pollicis. Patients may report their hand getting caught when putting their hand in a pocket. *Wartenberg's sign* is an abducted fifth digit at rest from relative weakness of the third palmar interosseous muscle, which adducts the fifth digit.

Froment's sign occurs when the patient is asked to make a pincher grip, as in gripping a bill between the thumb and index finger. Weakness of the adductor pollicis, deep head of the flexor pollicis brevis, and interossei will cause the patient to compensate with the median-innervated flexor pollicis longus and flexor digitorum profundus of the second digit. In a way, this is a reverse OK sign, which can be found in an anterior interosseous neuropathy.

Diagnostics

Electrodiagnostic studies are useful to differentiate an ulnar neuropathy from a C8 or T1 radiculopathy, medial cord, or lower trunk plexopathy, which can all present similarly. Non-ulnar-innervated muscles that receive C8–T1 innervation through the lower trunk and medial cord include abductor pollicis brevis (median) and longus (radial) and flexor pollicis longus (median) and brevis (recurrent branch of median).

Sensory exam of the medial antebrachial cutaneous nerve should be intact in an ulnar neuropathy since this nerve branches off the medial cord prior to its termination as the ulnar nerve. Sensory nerve study of the dorsal ulnar cutaneous sensory branch can localize the lesion to above the wrist if this nerve is abnormal since it branches off the ulnar nerve several centimeters proximal to the wrist. In addition, the palmar cutaneous sensory branch, which supplies sensation to the skin of the medial palmar surface of the hand, is given off a couple centimeters proximal to the wrist after the dorsal ulnar cutaneous nerve branches off. In a suspected ulnar neuropathy with sensory involvement in the hand, it is important to localize the lesion to the wrist in Guyon's canal or the elbow if surgery is being considered. Often, it is difficult to localize the lesion of an ulnar neuropathy, leaving the diagnosis as a nonlocalizable ulnar neuropathy [27].

Treatment

Ulnar Neuropathy at the Elbow

Nonoperative treatment can be considered in mild cases of ulnar neuropathy at the elbow. The natural course of untreated ulnar neuropathy at the elbow is favorable; approximately half of patients do recover spontaneously [28]. Activity modification such as avoiding pressure on the elbow or prolonged elbow flexion can be beneficial. Physical or occupational therapy for strengthening and maintaining range of motion and learning nerve gliding exercises, as well as elbow padding or splints, can be used to manage mild symptoms.

When there are signs and symptoms of more advanced disease such as motor weakness or atrophy, a surgical referral should be considered. Steroid injections at the elbow have not been shown to help this condition. A randomized controlled trial comparing elbow splinting to splinting plus steroid injection at the ulnar groove did not find a benefit with the addition of a steroid injection [29].

There are three surgical approaches for ulnar neuropathy at the elbow: in situ ulnar nerve release, ulnar release with medial epicondylectomy, and ulnar nerve release with anterior transposition. The latter two approaches have the additional goal of reducing the tensile stretch on the ulnar nerve, especially in cases where the ulnar nerve subluxes from the groove. A medial epicondylectomy will take the tension out of the ulnar nerve when it normally passes posterior to the epicondyle. The nerve transpositions involve physically moving the nerve from the ulnar groove and securing it anterior to the medial epicondyle of the humerus. Studies comparing in situ decompression to nerve transpositions have not found a significant difference in functional outcomes, but a review of the literature found that the nerve transpositions have a higher rate of medial antebrachial cutaneous neuropathy and postoperative infections [30].

Nonoperative treatment should be attempted in the absence of significant motor weakness or permanent sensory loss for ulnar neuropathy at the wrist. Activity modification and bracing can be beneficial, especially in individuals who have experienced repetitive insults to the ulnar nerve at the wrist. NSAIDs and steroid injections may be used in these cases.

Radial Neuropathy

Anatomy

The radial nerve receives contributions from C5 through T1. All three trunks of the brachial plexus give off posterior divisions that form the posterior cord. The posterior cord gives off the axillary, thoracodorsal, and subscapular nerves before terminating as the radial nerve.

In the proximal arm, the radial nerve gives off three sensory branches: *posterior cutaneous nerve of the arm*, *lower lateral cutaneous nerve of the arm*, and *posterior cutaneous nerve of the forearm*. It also supplies motor innervation to the three heads of the triceps brachii and the anconeus prior to entering the spiral groove in the humerus.

Near the elbow, the radial nerve supplies motor innervation to the brachioradialis, extensor carpi radialis longus, and the supinator. Several centimeters distal to the lateral epicondyle, the radial nerve splits into the superficial radial sensory nerve and the posterior interosseous nerve. The *superficial radial sensory branch nerve* innervates the lateral dorsal hand and thumb plus the dorsal proximal phalanges of digits 2, 3, and 4. The *posterior interosseous nerve* is clinically a pure motor nerve that dives beneath the supinator muscle under the arcade of Frohse. The posterior interosseous nerve supplies the following: extensor carpi radialis brevis, extensor digitorum communis, extensor carpi ulnaris, abductor pollicis longus, extensor indicis proprius, extensor pollicis longus, and extensor pollicis brevis.

Epidemiology

Radial neuropathy is less common than both median and ulnar neuropathies in the upper extremity. Radial neuropathy at the spiral groove is associated with humeral shaft fractures, particularly mid-shaft fractures. A review of over 4,500 humeral fractures found an 11.8 % incidence of radial nerve palsies with humeral fractures [31].

History and Clinical Presentation

Radial neuropathy at the spiral groove can be from compression secondary to prolonged immobilization as in the classic Saturday night palsy, when an intoxicated individual falls asleep with their arm draped over a chair. Another presentation is honeymooner's palsy, where the partner falls asleep with their head in the medial arm, compressing the radial nerve in the spiral groove. Other mechanisms include humeral fractures, after strenuous muscular effort (called exercise-induced radial neuropathy [32]), and vasculitic infarctions. Radial neuropathy at the spiral groove will present as wrist- and finger-drop with sparing of elbow extension since the muscular branches to the triceps brachii and anconeus muscles arise proximally to the spiral groove. There will be sensory disturbance in the hand in the radial distribution.

Involvement of the radial nerve in the axilla will present with similar symptoms as above with the additional impairment of elbow extension from triceps involvement and sensory impairment of the posterior arm and forearm from *posterior cutaneous nerve of the arm* and *lower lateral cutaneous nerve of the arm* involvement. In a *posterior interosseous neuropathy*, the patient will have wristdrop without sensory disturbance. Compared to a lesion at the spiral groove, there will be sparing of the strength of the brachioradialis and extensor carpi radialis longus, which may result in a radially deviated wrist in extension. *Radial tunnel syndrome* is a controversial disorder also known as resistant tennis elbow due to symptoms presenting near the lateral epicondyle. The cause of this syndrome is thought to be compression of the posterior interosseous nerve that results in pain in the proximal forearm extensor muscles without weakness [33].

In a *superficial radial sensory neuropathy*, there will be sensory disturbance of the lateral dorsal hand and thumb plus the dorsal proximal phalanges of digits 2, 3, and 4. The sensory nerve runs superficially along the distal third of the radius and can be prone to compression here. Tight watches, bracelets, or handcuffs could result in a superficial radial sensory neuropathy. Other diagnoses to consider presenting as weakness of the brachioradialis could be a C5–C6 radiculopathy rather than a radial neuropathy. Both a C7 radiculopathy and radial neuropathy at the axilla would affect the triceps strength and reflexes. A posterior brachial plexus cord lesion would affect the axillary and thoracodorsal nerves as well as the radial nerve.

Diagnostics

Radiographs of the elbow and forearm may be useful in evaluating anatomical structures that contribute to symptoms. Electrodiagnostic examination can help localize the lesion to the radial nerve. Radial motor nerve conduction studies may be especially useful for localization but are difficult to perform in males with larger shoulders.

Treatment

A radial nerve palsy following a closed humeral fracture can be initially observed for 6–12 weeks to allow swelling to subside. A review of outcomes for humeral fracture management showed that expectant management had similar outcomes to early surgical exploration, and there was spontaneous recovery in 71 % of patients treated nonsurgically with wrist splinting to address wristdrop [31]. External compression neuropathies of the radial nerve proper respond well to nonsurgical management. Regardless of the site of involvement, almost all external compressive lesions of the radial nerve result in neuropraxia and a reversible conduction block that improves with removal of the compressive force. Of note, radial neuropathy from a tight tourniquet can take up to a year to recover. In comparison, typical neuropraxic lesions take an average of 6 weeks to heal [34].

For both posterior interosseous neuropathy and radial tunnel syndrome, there are no randomized controlled trials looking at nonoperative versus surgical treatment. There is a retrospective review of fourteen cases of posterior interosseous neuropathy that recommends 6 months of nonoperative management prior to proceeding with surgery [35]. Another review demonstrates that outcomes from surgical intervention were generally favorable, with a large range of efficacy (67–95 %) [36]. Patient satisfaction with these results ranged from 40 to 83 %. Superficial radial sensory neuropathies are best treated nonoperatively. Local steroid injections and iontophoresis are treatments that are commonly used. A case report suggests the use of kinesiotaping as nonoperative treatment [37]. The favorable risk-to-benefit profile of this treatment may lead to more widespread use.

Suprascapular Neuropathy

Anatomy

The suprascapular nerve comes off the upper trunk of the brachial plexus proximally and contains C5 and C6 innervation. The nerve runs posteriorly under the trapezius and travels through the suprascapular notch of the scapula into the supraspinous fossa to innervate the supraspinatus muscle. The nerve supplies deep sensory fibers to the glenohumeral and acromioclavicular joints, but does not contain cutaneous sensory innervation. Continuing laterally, the suprascapular nerve wraps around the spinoglenoid notch of the scapular spine and passes into the infraspinous fossa to innervate the infraspinatus muscle.

Epidemiology

Repetitive overhead activities may predispose individuals to suprascapular neuropathies. In a study of high-level volleyball players, there was a 33 % prevalence of a suprascapular neuropathy, mostly at the suprascapular notch, based on electrodiagnostics [38]. In the literature, male athletes are the typical patient population reported on. However, it is not known if males are at an increased risk. A case series of 38 volleyball players found a nearly equal distribution among males and females (20 males and 18 females) [39].

There is an association between large rotator cuff tears and suprascapular neuropathies. In a retrospective study of 216 patients with rotator cuff tears, 26 were identified as having large tears with associated retraction of the tendon and fatty infiltration of the muscles. Seven of the twenty-six patients (38 %) had an associated suprascapular neuropathy.

History and Clinical Presentation

Suprascapular nerve entrapment usually occurs at the suprascapular notch, where the nerve passes under the transverse scapular ligament. The patient may report a deep shoulder pain along the superior aspect of the scapula. Clinically, this may present as weakness of shoulder abduction and/or external rotation. Involvement at the spinoglenoid notch is usually the result of a space-occupying lesion, like a cyst. Labral tears that extend from the glenohumeral joint can cause cysts in the spinoglenoid notch. Lipomas and other benign tumors, like ganglion cysts, can also cause entrapment [40]. Entrapment at the spinoglenoid notch would affect the infraspinatus only, since the supraspinatus muscle already received its innervation more proximally.

As mentioned above, repetitive overhead activities can predispose to a suprascapular neuropathy, particularly when there is shoulder abduction with scapular protraction. Volleyball, tennis, dancing, painting, pitching, and other overhand throwing are activities that are associated with an increased risk of suprascapular neuropathy.

Exam

On inspection, there may be atrophy of the supraspinatus and/or the infraspinatus. Infraspinatus atrophy may be easier to appreciate because the muscle is only partially covered by the trapezius. There is usually no cutaneous sensory disturbance in a suprascapular neuropathy. On motor exam, there may be weakness in shoulder abduction (supraspinatus) and/or external rotation (infraspinatus). However, other muscles contribute to these movements so the strength impairment may be subtle. Crossed body shoulder adduction on the affected side causes tightening of the spinoglenoid ligament and can exacerbate symptoms of an entrapment neuropathy at this location [41].

Diagnostics

MRI of the cervical spine may be useful to exclude a C5 or C6 radiculopathy, which can present similarly. MRI of the shoulder can be used to assess the rotator cuff as well as visualize the suprascapular and spinoglenoid notches. Cystic lesions and lipomas can be visualized with MRI.

Electrodiagnostics can be useful to localize the lesion to the suprascapular nerve and rule out a cervical radiculopathy or an upper trunk brachial plexopathy, which could present similarly. Needle EMG of the supraspinatus and infraspinatus can give information to help localize the lesion relative to the spinoglenoid notch. Patient reporting of pain scores after an injection of an anesthetic agent into the suprascapular notch can be used as a diagnostic tool.

Treatment

In the absence of an identifiable mass lesion, such as a ganglion cyst compressing the suprascapular nerve, a trial of nonoperative management should be attempted. Physical therapy is useful to maintain shoulder range of motion and strengthen the surrounding muscles to improve scapulothoracic mechanics [42]. Injection of an anesthetic agent and/or corticosteroid into the suprascapular notch can be used to treat symptoms.

In the cases where nonoperative management does not lead to improvement or if there is a physical structure compressing the suprascapular nerve, surgical referral is indicated. For involvement at the suprascapular notch, release of the transverse scapular ligament may be beneficial. This can be done with either an open or arthroscopic approach [43].

Scapular Winging

Scapular winging is a condition of dysfunction of the muscles involved in scapulothoracic motion. Scapular winging can result in significant impairment in the use of the affected upper extremity. The altered mechanics of the scapula depend on which muscle is involved: serratus anterior, trapezius, or rhomboids. If the altered scapulothoracic mechanics are from local muscle dysfunction without nerve injury, this is called scapular dyskinesia. True nerve injury results in scapular winging. Long thoracic nerve palsy will result in a medial translation of the scapula due to unopposed action of the intact trapezius and rhomboids, known as medial winging. In contrast, individual palsies of the spinal accessory or dorsal scapular nerves will cause a lateral winging of the scapula. We will review the individual muscles implicated and the relevant anatomy, clinical presentation, and treatment options.

Long Thoracic Nerve

The long thoracic nerve is a motor nerve made by direct branches from the C5 to C7 nerve roots. The nerve courses under the clavicle and over the anterolateral rib cage to innervate the serratus anterior. This muscle has its origin on the first eight to ten thoracic ribs and inserts on the underside of the medial border of the scapula. The action is to keep the medial border of the scapula against the ribcage as it assists in scapular protraction.

Clinically, long thoracic nerve dysfunction is commonly the result of a more global injury to the cervical nerve roots or brachial plexus. For example, long thoracic nerve palsy can happen in a traumatic root avulsion or neuralgic amyotrophy, rather than an isolated peripheral nerve palsy [44]. However, compression can occur over the rib cage, such as in individuals carrying heavy backpacks, or by the middle scalene in individuals who lift heavy weights. Traction injuries can happen with abrupt forceful shoulder movements as in shoveling snow, weightlifting, or wrestling because the nerve is anchored prior to entering the axilla [45].

Since the serratus anterior is one of the shoulder-stabilizing muscles, the patient may report vague shoulder pain. Weakness of the serratus anterior causes medial winging of the scapula, which causes the entire scapula to be displaced medial and superior compared to the unaffected side. This can be more pronounced when the patient attempts to push against a wall. Sensory impairment suggests a lesion outside the long thoracic nerve since there is no cutaneous sensory information carried in the nerve.

Electrodiagnostics can differentiate a long thoracic nerve palsy from a cervical radiculopathy or brachial plexopathy. Motor nerve conduction study of the long thoracic nerve is technically challenging. Needle EMG study is the most helpful diagnostic test of the serratus anterior but should be performed with caution due to risk of an iatrogenic pneumothorax.

Nonoperative management can be successful as long as there is no root avulsion. In a retrospective study of 37 patients with mild long thoracic nerve palsy treated nonoperatively and without functional bracing, 78 % of the patients had resolution of their winging after an average of 16 months. However, there was residual pain in 70 % of the individuals on follow-up [46]. Physical therapy includes a focus on improving the mechanics of scapular motion as well as strengthening of the rotator cuff muscles. Analgesics may be used for comfort [47].

The traumatic and iatrogenic causes of long thoracic neuropathy are usually treated surgically if the impairment persists and significantly interferes with function. There are three categories of procedures: scapulothoracic fusion, static stabilization, and dynamic muscle transfer. Dynamic muscle transfer is considered the most effective with transfer of the sternal head of the pectoralis major to the inferior angle of the scapula being the preferred procedure [48].

Spinal Accessory Nerve (CN XI)

The spinal accessory nerve, also known as cranial nerve XI, is composed of two components. The spinal part is formed from contributions from the ventral horn of C1 through C4 cervical segments and travels along the lateral aspect of the spinal cord to enter the skull through the foramen magnum. It is joined by the accessory contribution originating in the nucleus accumbens of the medulla prior to exiting the skull from the jugular foramen. The spinal nerve then divides back into the accessory branch, which joins the vagus nerve, and the spinal branch, which is the pure motor branch that innervates the sternocleidomastoid (SCM) and trapezius muscles.

Clinically, spinal accessory nerve palsy can occur from compression in the posterior cervical triangle of the neck. This is a region of the neck formed by the posterior border of the SCM, anterior border of the trapezius, and the middle third of the clavicle that forms the base. Surgical procedures in the neck can inadvertently affect the distal spinal accessory nerve here. In fact, iatrogenic injury to the spinal accessory nerve during cervical lymph node biopsy or benign tumor removal is the most common cause of spinal accessory nerve palsy. Since the trapezius is a shoulder-stabilizing muscle, the patient often reports shoulder pain. Functionally, the patient often reports a drooping shoulder and difficulty with shoulder abduction [49].

On physical exam, the patient may have weakness of shoulder shrug of the affected side. Commonly, they will not be able to abduct the shoulder past 90°. There will be lateral winging of the scapula with the superior border of the scapula translated laterally. Winging of the scapula due to trapezius palsy should decrease with resisted forward flexion of the shoulder, where it would persist if the serratus anterior was the affected muscle. An interesting finding in a study of 83 patients presenting with spinal accessory nerve palsy was that weakness of the sternocleidomastoid was found in only one of the 83 individuals [49].

Unfortunately, nonsurgical treatment of a spinal accessory nerve palsy does not have the same positive outcomes compared to a serratus palsy. In fact, physical therapy, transcutaneous nerve stimulation, bracing, chiropractic manipulation, NSAIDS, and narcotic analgesics all were found to be ineffective [50]. Regardless, these are often attempted given the favorable risk-to-benefit profile. In neurapraxic lesions that do not improve after 1 year, surgical correction may be indicated [51]. The Eden-Lange procedure is preferred, which involves the lateral transfer of the insertion sites of the rhomboids and levator scapulae along the scapula to balance the forces of these intact muscles. Outcomes for this procedure are favorable, with one study showing that only 3 of 22 patients undergoing the procedure had unfavorable results with persistent pain and lack of functional improvement [52].

Dorsal Scapular Nerve

The dorsal scapular nerve is a terminal nerve that is given off from the C5 ventral rami prior to it forming the upper trunk with C6. In a few individuals, C4 or C6 can contribute to the dorsal scapular nerve [53, 54]. The dorsal scapular nerve pierces the scalenus medius prior to innervating the levator scapula and the rhomboid major and minor muscles. Together, the two rhomboid muscles act to retract and elevate the scapula as well as rotate its lateral border inferiorly. Dorsal scapular nerve palsies are usually due to compression, although the actual site of involvement is not well known [55].

In a cohort of 36 patients with dorsal scapular nerve compression, the most common presenting complaint was discomfort of the neck, back, and shoulder, particularly down the medial aspect of the scapula [55]. Of note, females made up 78 % of the group, but it was unclear what factor accounted for the difference in distribution.

Physical exam may reveal mild lateral translation of the inferior border of the scapula. Strength of the rhomboids can be assessed by having the patient squeeze the medial borders of the scapula together or having them put their hands on their hips and resist an anterior force applied to the back of their elbow.

Nonsurgical management of dorsal scapular nerve palsies is the mainstay of treatment. Cervical bracing and analgesics can be used to address the painful symptoms. Physical therapy will focus on strengthening the midportion of the trapezius, which can compensate for weak rhomboids. Surgical decompression of the anterior and middle scalene muscles can be done in individuals with symptoms refractory to therapy and medication management. Nineteen of a cohort of twenty-two patients with persistent symptoms despite nonsurgical management had partial or complete relief of their symptoms after decompression of the scalene muscles [55].

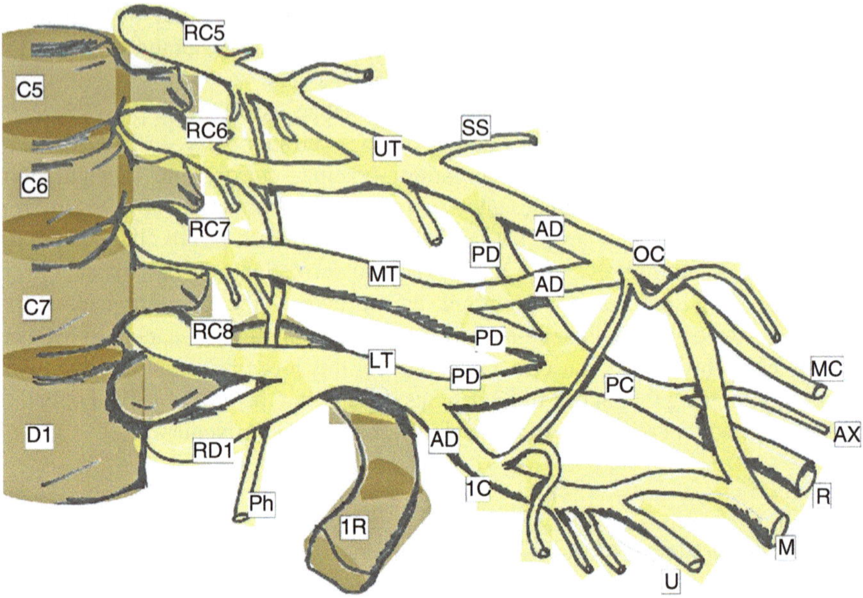

Fig. 4.2 Schematic anatomy of the brachial plexus. Bone structures (C5–C6–C7–D1 fifth to seventh cervical spine vertebrae and first dorsal spine vertebrae, 1R first rib). Brachial plexus components (RC5–RC8 fifth to eighth cervical brachial plexus roots, RD1 firstst dorsal brachial plexus root, *UT* upper trunk, *MT* middle trunk, *LT* lower trunk, *PD* posterior division from each trunk, *AD* anterior division from each trunk, *OC* outer chord, *PC* posterior chord, *IC* inner chord). Main nerve branches (*Ph* phrenic nerve, *SS* suprascapular nerve, *MC* musculocutaneous nerve, *Ax* axillary nerve, *R* radial nerve, *M* median nerve, *U* ulnar nerve (Used with permission of Springer Science+Business Media from Ruggiero and Liverneaux [66]))

Brachial Plexopathy

Anatomy

The brachial plexus is a structure formed by the ventral rami of the lower cervical and upper thoracic nerve roots. In the majority of cases, the brachial plexus is formed from the C5 through T1 levels, although anatomical variations do exist. In some individuals, levels C4 through C7 form the major contributions to the brachial plexus; this is termed *prefixed*. In others, levels C6 through T2 form the major contributions; this is called *postfixed*. The prevalence of these anatomical "variations" is more common than one may realize. A surgical review of multiple anatomical studies showed that the prevalence of a prefixed brachial plexus is 34 % and a postfixed plexus is 19 % [56].

Physically, the brachial plexus is located between the anterior and middle scalene muscles proximally and posterior to the clavicle and pectoral muscles distally. The axillary artery is in close proximity to the brachial plexus. The brachial plexus is divided into roots, trunks, divisions, cords, and nerves. See Fig. 4.2 for schematic of the anatomy of the brachial plexus. Classically, the levels C5 through T1 contribute their ventral rami to form the roots of the plexus. The *upper trunk* is formed by C5 and C6, the *middle trunk* by C7, and the *lower trunk* by C8 and T1. Each trunk then divides into an *anterior and a posterior division*. The posterior divisions of the upper, middle, and lower trunks come together to form the *posterior cord*. The anterior divisions of the upper and middle trunk form the *lateral cord* and the anterior division of the lower trunk continues as the *medial cord*. Most of the major nerves of the upper extremity come off the cords; the dorsal scapular and long thoracic nerves originate directly from the nerve roots and the suprascapular nerve originates from the upper trunk [57].

Epidemiology

Traumatic injuries to the brachial plexus are the most common etiology of brachial plexopathies. The plexus may be affected in motor vehicle accidents, bicycle accidents, or penetrating knife or gunshot wounds. Iatrogenic injuries to the brachial plexus can occur during coronary artery bypass or other thoracic procedures as a result of trauma to the plexus during chest wall retraction. Another mechanism is a traction injury to the newborn during delivery.

Besides traumatic injuries, the brachial plexus can be compressed mechanically. Lymph node metastasis from malignancy can cause a mass effect and compress the brachial plexus. Breast cancer, lymphomas, and lung cancer are the most frequent cause of lymphadenopathy causing plexopathy. Local tumor invasion may occur as well. This is seen in Pancoast tumors of the lung, lymphomas, and leukemia. Less common are the primary tumors involving the nerve sheath, such as schwannomas, neurofibromas, and neurofibrosarcomas. Nonneoplastic mass effects may occur, such as hematomas from internal jugular catheters or vascular abnormalities like arteriovenous malformations or aneurysms [58]. There is also true neurogenic thoracic outlet syndrome, which is usually the result of a fibrous band between a cervical rib and the first thoracic rib that compresses the lower trunk of the brachial plexus.

Radiation-induced brachial plexitis is a progressive plexopathy that presents typically years after the initial radiation exposure. Radiation treatment protocols often include the region of the brachial plexus, especially for treatment of lymphomas and breast, lung, and neck cancers. The differential diagnosis for neurologic symptoms in a patient with history of malignancy treated with radiation is either a radiation neuropathy or recurrent tumor invading the nerves. Sensory symptoms (dysesthesias and numbness) are more common than pain in radiation plexopathies compared to plexopathies from direct tumor invasion, where pain is the most prominent complaint [58].

Neuralgic amyotrophy (aka brachial plexitis, Parsonage-Turner syndrome, brachial amyotrophy) exists in two forms: a hereditary form and an idiopathic form. The hereditary form is an autosomal-dominant condition characterized by repeated episodes of upper extremity neuropathy. The pathophysiology is thought to be an immune-mediated attack on the brachial plexus. Commonly, this is preceded by a viral illness, immunization, childbirth, or other stressor on the immune system. The annual incidence of neuralgic amyotrophy is 2–3 per 100,000 persons per year and affects men more than women at a ratio of 3:2 [59].

History and Clinical Presentation

The presentation of a brachial plexopathy depends on the portion of the brachial plexus that is involved [58]:

- *Panplexopathy*: involvement of the entire brachial plexus will result in weakness, sensory loss, and decreased reflexes in the entire arm. If the nerve roots are intact, the serratus anterior and the rhomboids are the only muscles spared because the nerves that innervate these muscles come directly from the nerve roots.
- *Upper trunk plexopathy*: the upper trunk is formed by the C5 and C6 nerve roots so there may be prominent weakness in the deltoid, biceps, brachioradialis, supraspinatus, and infraspinatus. Other muscles that receive contributions from C5 and C6, such as the pronator teres and triceps, may be partially affected. Sensory loss will involve the lateral arm, lateral forearm, lateral hand, and the thumb. The biceps and brachioradialis reflexes will be depressed. Clinically, upper plexopathies can occur from injuries where the head is pushed forcefully away from the shoulder, causing traction on the upper root fibers. This presents as the classic Erb's palsy: weakness of shoulder abduction, elbow flexion, and arm supination.
- *Middle trunk plexopathy*: presents similarly to a C7 radiculopathy since the middle trunk is formed by the C7 root. Weakness may be found in the triceps, flexor carpi radialis, and pronator teres. Sensory loss will involve the middle finger and may include the index and ring fingers (sensory of median nerve) and posterior forearm (posterior cutaneous nerve of forearm). The triceps reflex will be depressed or absent.
- *Lower trunk plexopathy*: the lower trunk is formed by the C8 and T1 nerve roots so there may be weakness in all the ulnar-innervated muscles (intrinsic hand muscles) as well as the median C8–T1-innervated muscles (abductor pollicis brevis, flexor pollicis longus, and flexor digitorum profundus) and radial C8-innervated muscles (extensor indicis proprius and extensor pollicis brevis). Sensory loss will involve the medial arm, medial forearm, medial hand, and the fifth digit. Reflexes will be preserved. Clinically, lower plexopathies can occur when the arm and shoulder are forcefully pulled up above the head. This presents as the classic Klumpke's palsy: weakness of the intrinsic hand muscles and preservation of the proximal shoulder and arm muscles.
- *Lateral cord plexopathy*: the musculocutaneous nerve and the C6–C7 portion of the median nerve will be affected, resulting in weakness of arm pronation (pronator teres), wrist flexion (flexor carpi radialis), and elbow flexion (biceps). Sensory loss will involve the lateral forearm, lateral hand, and first three fingers.

The biceps reflex will be affected, while the triceps and brachioradialis reflexes are preserved.

- *Medial cord plexopathy*: the anterior division of the lower trunk becomes the medial cord, so medial cord plexopathies present similarly to lower trunk plexopathies with the exception of intact C8 radial fibers, which pass through the posterior division of the lower trunk and posterior cord. Preservation of the finger extensors (extensor indicis proprius and extensor pollicis brevis) will differentiate a medial cord lesion from a lower trunk lesion.
- *Posterior cord plexopathy*: the radial, axillary, thoracodorsal, and subscapular nerves all branch off the posterior cord. A posterior cord plexopathy will result in weakness in elbow extension, wristdrop, and fingerdrop from radial involvement; weakness in shoulder abduction from deltoid involvement; and weakness in shoulder adduction from latissimus dorsi involvement. Sensory loss will involve the lateral arm (axillary nerve), posterior arm and forearm (radial nerve), and the radial aspect of the dorsum of the hand (radial nerve). The triceps and brachioradialis reflexes will be depressed.
- *Neuralgic amyotrophy*: this can be preceded by a viral illness, immunization, childbirth, or other stressor on the immune system. Initially, there is shoulder pain that is constant and lasts approximately 4 weeks. If severe, this pain may inhibit functional strength. Neurologic weakness becomes apparent after 1–2 weeks. The pattern of symptoms is usually unilateral in the upper and middle trunk distribution with the long thoracic and suprascapular nerves being the most commonly affected [60].

Diagnostics

Electrodiagnostic testing is valuable to localize the lesion and assess the severity of the brachial plexopathies. There are two parts to the electrodiagnostic exam, the nerve conduction studies and the needle exam. The nerve conduction studies assess the speed of conduction in the sensory and motor nerves. Since sensory studies are normal in radiculopathy, the presence of abnormal sensory studies in conjunction with an abnormal needle exam in the distribution of the patient's complaints is diagnostic of brachial plexopathy (in the absence of a more distal compressive lesion, such as median neuropathy at the wrist or ulnar neuropathy at the elbow.)

Treatment

Traumatic injuries to the brachial plexus where the roots are avulsed from the spinal cord carry a poor prognosis for recovery. In general, the upper trunk lesions are more favorable since there is residual hand function. Repair should be done as soon as possible, ideally within days, to maximize recovery. Repair can be done by either nerve transfer or nerve grafts [61].

Treatment of radiation-induced brachial plexopathy involves systemic steroids, anticoagulation, and hyperbaric oxygen treatment [62].

There is no accepted standard of care treatment for neuralgic amyotrophy. Corticosteroids are commonly prescribed. Van Ejik et al. did a retrospective analysis of 50 patients with neuralgic amyotrophy treated with steroids compared to a historical cohort of 203 untreated patients. Median time to a decrease in pain intensity was 12.5 days in the steroid group compared to 20.5 days in the untreated group. Recovery of strength was seen by the first month in 18 % of the steroid group versus 6.3 % of the untreated group and after 1 year. Full recovery was found in 12 % of the steroid group versus 1 % in the untreated group. Side effects were found in 20 % of the treatment group, but did not lead to discontinuation of the steroids [63]. Intravenous immunoglobulin with pulsed steroids has been used as treatment with good results. However, there is no trial to study the efficacy of this treatment against others [64].

Conclusion

Median and ulnar neuropathy are by far the most common upper limb entrapment syndromes. Ulnar neuropathy may be more common in males, but further data is needed to determine this. The practitioner should be aware of the potential for brachial plexopathy. However rare, this can happen during childbirth, which is unique to the female. With the exception of median neuropathy at the wrist, treatment is largely supportive.

References

1. Preston DC, Shapiro BE. Electromyography and neuromuscular disorders, chapters 17. 2nd ed. Philadelphia: Elsevier; 2005.
2. Stevens JC, Sun S, Beard CM, O'Fallon WM, Kurland LT. Carpal tunnel syndrome in Rochester, Minnesota, 1961 to 1980. Neurology. 1988;38(1):134–8.
3. Stevens JC, Beard CM, O'Fallon WM, Kurland LT. Conditions associated with carpal tunnel syndrome. Mayo Clin Proc. 1992;67(6):541–8.
4. Werner RA, Albers JW, Franzblau A, Armstrong TJ. The relationship between body mass index and the diagnosis of carpal tunnel syndrome. Muscle Nerve. 1994;17(6):632–6.
5. Solomon DH, Katz JN, Bohn R, Mogun H, Avorn J. Nonoccupational risk factors for carpal tunnel syndrome. J Gen Intern Med. 1999;14(5):310–4.
6. Palmer KT, Harris EC, Coggon D. Carpal tunnel syndrome and its relation to occupation: a systematic literature review. Occup Med (Lond). 2007;57(1):57–66.
7. Thomsen JF, Gerr F, Atroshi I. Carpal tunnel syndrome and the use of computer mouse and keyboard: a systematic review. BMC Musculoskelet Disord. 2008;9:134.
8. Stevens JC, Smith BE, Weaver AL, Bosch EP, Deen Jr HG, Wilkens JA. Symptoms of 100 patients with electromyographically verified carpal tunnel syndrome. Muscle Nerve. 1999;22(10):1448–56.
9. DiBenedetto M, Giering R. Chapter 55: Upper extremity nerve entrapment. In: O'Connor F, editor. Sports medicine just the facts (327). New York: McGraw-Hill Companies; 2005.

10. MacDermid JC, Wessel J. Clinical diagnosis of carpal tunnel syndrome: a systematic review. J Hand Ther. 2004;17(2):309–19.
11. Preston DC, Shapiro BE. Electromyography and neuromuscular disorders, chapters 18. 2nd ed. Philadelphia: Elsevier; 2005.
12. Rempel D, Tittiranonda P, Burastero S, Hudes M, So Y. Effect of keyboard keyswitch design on hand pain. J Occup Environ Med. 1999;41:111–9.
13. Piazzini DB, Aprile I, Ferrara PE, Bertolini C, Tonali P, Maggi L, Rabini A, Piantelli S, Padua L. A systematic review of conservative treatment of carpal tunnel syndrome. Clin Rehabil. 2007;21(4):299–314.
14. Manente G, Torrieri F, Di Blasio F, Staniscia T, Romano F, Uncini A. An innovative hand brace for carpal tunnel syndrome: a randomized controlled trial. Muscle Nerve. 2001;24:1020–5.
15. Walker WC, Metzler M, Cifu DX, Swartz Z. Neutral wrist splinting in carpal tunnel syndrome: a comparison of night-only versus full-time wear instructions. Arch Phys Med Rehabil. 2000;81:424–9.
16. Hui AC, Wong SM, Wong KS, et al. Oral steroid in the treatment of carpal tunnel syndrome. Ann Rheum Dis. 2001;60:813–4.
17. O'Connor D, Marshall S, Massy-Westropp N. Non-surgical treatment (other than steroid injection) for carpal tunnel syndrome. Cochrane Database Syst Rev. 2003;1, CD003219.
18. Marshall S, Tardif G, Ashworth N. Local corticosteroid injection for carpal tunnel syndrome. Cochrane Database Syst Rev. 2007;2, CD001554.
19. Huisstede BM, Randsdorp MS, Coert JH, Glerum S, van Middelkoop M, Koes BW. Carpal tunnel syndrome. Part II: effectiveness of surgical treatments–a systematic review. Arch Phys Med Rehabil. 2010;91(7):1005–24.
20. Eversmann WW. Proximal median nerve compression. Hand Clin. 1992;8(2):307–15.
21. Elhassan B, Steinmann SP. Entrapment neuropathy of the ulnar nerve. J Am Acad Orthop Surg. 2007;15(11):672–81.
22. Mondelli M, Giannini F, Ballerini M, Gianneschi F, Martorelli E. Incidence of ulnar neuropathy at the elbow in the province of Siena (Italy). J Neurol Sci. 2005;234(1–2):5–10.
23. Lederman RJ. Focal peripheral neuropathies in instrumental musicians. Phys Med Rehabil Clin N Am. 2006;17(4):761–79.
24. Contreras MG, et al. Anatomy of the ulnar nerve at the elbow: potential relationship of acute ulnar neuropathy to gender differences. Clin Anat. 1998;11(6):372–8.
25. DiBenedetto M, Giering R. Chapter 55: Upper extremity nerve entrapment. In: O'Connor F, editor. Sports medicine just the facts (328). New York: McGraw-Hill Companies; 2005.
26. Gelberman RH, Yamaguchi K, Hollstein SB, et al. Changes in interstitial pressure and cross-sectional area of the cubital tunnel and of the ulnar nerve with flexion of the elbow: an experimental study in human cadavera. J Bone Joint Surg Am. 1998;80:492–501.
27. Preston DC, Shapiro BE. Electromyography and neuromuscular disorders, chapter 19. 2nd ed. Philadelphia: Elsevier; 2005.
28. Robert M. Szabo, Christine Kwak, natural history and conservative management of cubital tunnel syndrome. Hand Clin. 2007;23(3):311–8.
29. Hong CZ, Long HA, Kanakamedala RV, Chang YM, Yates L. Splinting and local steroid injection for the treatment of ulnar neuropathy at the elbow: clinical and electrophysiological evaluation. Arch Phys Med Rehabil. 1996;77(6):573–7.
30. McPherson SA, Meals RA. Cubital tunnel syndrome. Orthop Clin North Am. 1992;23(1):111–23.
31. Shao YC, Harwood P, Grotz MR, Limb D, Giannoudis PV. Radial nerve palsy associated with fractures of the shaft of the humerus: a systematic review. J Bone Joint Surg Br. 2005; 87(12):1647–52.
32. Nukada H, Taylor PK, August SD. Isolated triceps weakness in exercise-induced radial neuropathy. J Sports Med Phys Fitness. 1996;36:287.
33. Ritts GD, Wood MB, Linscheid RL. Radial tunnel syndrome. A ten-year surgical experience. Clin Orthop. 1987;219:201–5.
34. Wilbourn AJ. Nerve injuries caused by injections and tourniquets. AAEM Plenary Session: Physical Trauma to Peripheral Nerves, Minneapolis, MN, 1996. p. 229.

35. Vrieling C, Robinson PH, Geertzen JH. Posterior interosseous nerve syndrome: literature review and report of 14 cases. Eur J Plast Surg. 1998;21:196–202.
36. Huisstede B, Miedema HS, van Opstal T, et al. Interventions for treating the radial tunnel syndrome: a systematic review of observational studies. J Hand Surg Am. 2008;33:72–8.
37. Anandkumar S. Kinesio tape management for superficial radial nerve entrapment: a case report. Physiother Theory Pract. 2013;29(3):232–41.
38. Holzgraefe M, Kukowski B, Eggert S. Prevalence of latent and manifest suprascapular neuropathy in high-performance volleyball players. Br J Sports Med. 1994;28(3):177–9.
39. Ferretti A, De Carli A, Fontana M. Injury of the suprascapular nerve at the spinoglenoid notch. The natural history of infraspinatus atrophy in volleyball players. Am J Sports Med. 1998; 26(6):759–63.
40. Fehrman DA, Orwin JF, Jennings RM. Suprascapular nerve entrapment by ganglion cysts: a report of six cases with arthroscopic findings and review of the literature. Arthroscopy. 1995; 11(6):727–34.
41. Plancher KD, Luke TA, Peterson RK, Yacoubian SV. Posterior shoulder pain: a dynamic study of the spinoglenoid ligament and treatment with arthroscopic release of the scapular tunnel. Arthroscopy. 2007;23(9):991–8.
42. Martin SD, Warren RF, Martin TL, Kennedy K, O'Brien SJ, Wickiewicz TL. Suprascapular neuropathy. Results of non-operative treatment. J Bone Joint Surg Am. 1997;79(8):1159–65.
43. Freehill MT, Shi LL, Tompson JD, Warner JJ. Suprascapular neuropathy: diagnosis and management. Phys Sportsmed. 2012;40(1):72–83.
44. Wilbourn A, Ferrante M. Chapter 60: Upper limb neuropathies. In: Dyck P, Thomas PK, editors. Peripheral neuropathy. 4th ed. Philadelphia: Saunders; 2005. p. 1463–86.
45. Vastamäki M, Kauppila LI. Etiologic factors in isolated paralysis of the serratus anterior muscle: a report of 197 cases. J Shoulder Elbow Surg. 1993;2(5):240–3.
46. Pikkarainen V, Kettunen J, Vastamäki M. The natural course of serratus palsy at 2 to 31 years. Clin Orthop Relat Res. 2013;471(5):1555–63.
47. DiBenedetto M, Giering R. Chapter 55: Upper extremity nerve entrapment. In: O'Connor F, editor. Sports medicine just the facts (324). New York: McGraw-Hill Companies; 2005.
48. Warner JJ, Navarro RA. Serratus anterior dysfunction. Recognition and treatment. Clin Orthop Relat Res. 1998;349:139–48.
49. Donner TR, Kline DG. Extracranial spinal accessory nerve injury. Neurosurgery. 1993;32(6): 907–10.
50. Bigliani LU, Perez-Sanz JR, Wolfe IN. Treatment of trapezius palsy. J Bone Joint Surg Am. 1985;67: 871–7.
51. Wiater JM, Bigliani LU. Spinal accessory nerve injury. Clin Orthop Relat Res. 1999;368: 5–16.
52. Bigliani LU, Compito CA, Duralde XA, Wolfe IN. Transfer of the levator scapulae, rhomboid major, and rhomboid minor for paralysis of the trapezius. J Bone Joint Surg Am. 1996;78(10): 1534–40.
53. Tubbs RS, Tyler-Kabara EC, Aikens AC, Martin JP, Weed LL, Salter EG, Oakes WJ. Surgical anatomy of the dorsal scapular nerve. J Neurosurg. 2005;102:910–1.
54. Amato AA, Russell JA. Neuromuscular disorders, chapter 21. New York: The McGraw-Hill Company; 2008.
55. Desong C, Yudong G, Jie L, Liang C. Dorsal scapular nerve compression atypical thoracic outlet syndrome. Chin Med J (Engl). 1995;108:582–5.
56. Pellerin M, Kimball Z, Tubbs RS, Nguyen S, Matusz P, Cohen-Gadol AA, Loukas M. The prefixed and postfixed brachial plexus: a review with surgical implications. Surg Radiol Anat. 2010;32(3):251–60.
57. Jenkins DB. Hollinshead's functional anatomy of the limbs and back, chapter 4. 8th ed. Philadelphia: W.B. Saunders Company; 2002.
58. Preston DC, Shapiro BE. Electromyography and neuromuscular disorders, chapter 30. 2nd ed. Philadelphia: Elsevier; 2005.

59. Stutz CM. Neuralgic amyotrophy: Parsonage-Turner syndrome. J Hand Surg Am. 2010;35(12): 2104–6.
60. van Alfen N, van Engelen BG. The clinical spectrum of neuralgic amyotrophy in 246 cases. Brain. 2006;129:438–50.
61. Birch R. Chapter 62: Operating on peripheral nerves. In: Dyck P, Thomas PK, editors. Peripheral neuropathy. 4th ed. Philadelphia: Saunders; 2005. p. 1511–32.
62. Zhu J. Recent progress in diagnosis and treatment of the injury to the peripheral nerve. Article in Chinese. Zhongguo Xiu Fu Chong Jian Wai Ke Za Zhi. 2006;20(4):319–23.
63. van Eijk JJ, van Alfen N, Berrevoets M, van der Wilt GJ, Pillen S, van Engelen BG. Evaluation of prednisone treatment in the acute phase of neuralgic amyotrophy: an observational study. J Neurol Neurosurg Psychiatry. 2009;80:1120–4.
64. Naito KS, Fukushima K, Suzuki S, Kuwahara M, Morita H, Kusunoki S, Ikeda S. Intravenous immunoglobulin with methylprednisolone pulse therapy for motor impairment of neuralgic amyotrophy: clinical observations in 10 cases. Intern Med. 2012;51(12):1493–500.
65. Kang JR, Gupta R. Reoperative hand surgery. In: Duncan SFM, editor. Reoperative options for compressive neuropathies of the upper extremity. New York: Springer Science+Business Media; 2012.
66. Ruggiero GM, Liverneaux P. Brachial plexus repair. In: Liverneaux PA, Berner SH, Bednar MS, Parekattil SJ, Ruggiero GM, Selber JC, editors. Telemicrosurgery. New York: Springer Science+Business Media; 2013.

Chapter 5
Lower Limb Nerve Entrapment Syndromes

Kentaro Onishi and Jennifer Baima

Abstract The largest nerve in the body, the sciatic nerve, divides into the peroneal and tibial nerves proximal to the popliteal fossa. Patients with peroneal nerve injury present with foot drop, frequent falls, and dysesthesias over lateral half of anterior leg and/or dorsum of the foot. Gender-related concerns in peroneal neuropathy include prolonged knee flexion during parturition, female prevalence of gastric bypass, and female prevalence of ligamentous knee injuries. Patients with tibial nerve injury present with dysesthesias over medial malleolus and/or sole of the foot including the toenails, nocturnal exacerbation of symptoms, and motor weakness in plantar flexion and/or inversion of ankle. Gender issues in tibial neuropathy include prevalence of rheumatologic disease and ligamentous injuries in women.

Nerve lesions can afflict the upper or lower lumbar plexus. Patients with upper lumbar plexopathies present any combination of the following: weakness in hip flexion, knee extension, and hip abduction; sensory abnormality in anterior pelvic area and anterior, medial, or lateral thigh with radiation down to posterior-medial leg; and absent or decreased deep tendon reflex at the patellar tendon. Patients with lower lumbosacral plexopathies present with any combination of the following: weakness in hip extension, hip abduction, knee flexion, and motion at the ankle or

K. Onishi, DO
Department of Physical Medicine and Rehabilitation,
University of California, Irvine Medical Center,
Orange, CA 92868, USA
e-mail: kenonishi918@gmail.com

J. Baima, MD (✉)
Division of Women's Health, Department of Medicine,
Harvard Medical School, Brigham and Women's Hospital,
Chestnut Hill, MA 02467, USA
e-mail: jbaima1@partners.org

E. Mody, E. Matzkin (eds.), *Musculoskeletal Health in Women*,
DOI 10.1007/978-1-4471-4712-1_5, © Springer-Verlag London 2014

toes; sensory abnormality in the posterior thigh, all of the leg except for saphenous nerve area, and both dorsum and sole of the foot; and absent or decreased deep tendon reflex at the Achilles tendon. Gender issues in plexopathy include parturition-related injuries and breast and gynecologic cancers.

Keywords Peroneal neuropathy • Foot drop • Tibial neuropathy • Lumbosacral plexopathy

Peroneal Nerve

Anatomy of the Peroneal Nerve

The peroneal nerve arises from the L4–S1 nerve roots. These nerve roots merge with nerve fibers of tibial nerve to form the largest peripheral nerve in the human body, the sciatic nerve. The sciatic nerve exits through greater sciatic notch into gluteal area. The nerve comes under the piriformis muscle as it makes its way inferiorly to posterior thigh. The sciatic nerve then gives off a number of muscular branches to the hamstring muscles as it courses down posterior-medial thigh along with long head of biceps femoris. Then, it divides into the common peroneal and the tibial nerves proximal to popliteal fossa.

The lateral portion of the sciatic nerve innervates the short head of the biceps femoris proximal to the popliteal fossa, but this will become the common peroneal nerve as the nerve divides in the distal third of the thigh. This is the most distal sciatic-innervated muscle. In some individuals, a sensory branch called the lateral sural cutaneous nerve courses inferolaterally along the tendon of the same muscle. The peroneal nerve then sends out the lateral cutaneous nerve of the knee before it wraps around the lateral aspect of the fibular head. See Fig. 5.1 for schematic of the location of the peroneal nerve at the knee and its relationship to the peroneus longus tendon. Due to its proximity to a hard bony structure (fibular head) and to its superficial location, the fibular neck is the most common site of peroneal injury.

As the nerve runs lateral to fibular head, it passes between the fibular head and peroneus longus tendon. This space is known as "fibular tunnel" and is another site for entrapment nerve injury. Just distal to the tunnel is the division of the common peroneal nerve to both deep and superficial peroneal nerves. The deep peroneal nerve courses inferomedially to innervate dorsiflexors of the ankles and extensors of toes, while the superficial peroneal nerve stays over on the anterolateral aspect of the leg to innervate everters of the ankle.

The deep peroneal nerve is also responsible for cutaneous innervation to thefirst web space, while superficial peroneal nerve supplies the sensation to the middle and lower lateral calf. The superficial peroneal nerve further divides to medial and intermediate dorsal cutaneous nerves of the foot proximal to the ankle joint, innervating sensation to dorsum of the foot [1].

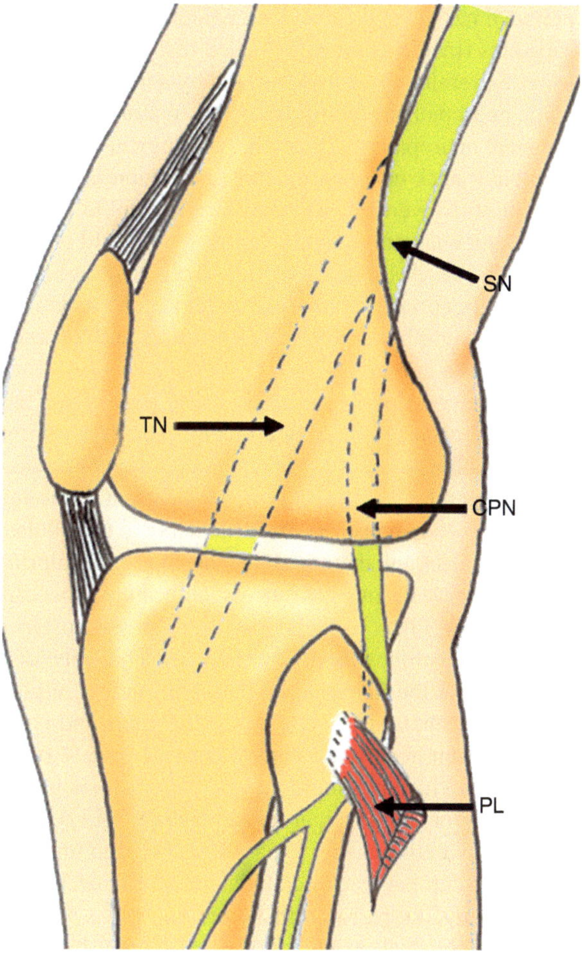

Fig. 5.1 Schematic diagram of the course of the common peroneal nerve. The sciatic nerve (*SN*) branches into the tibial nerve (*TN*) and common peroneal nerve (*CPN*) at the level of the proximal popliteal fossa. The common peroneal nerve winds around the fibular head, passes below the peroneus longus tendon (*PL*), and then branches into the superficial and deep branches (Used with permission from Kim et al. [66])

Symptoms and Signs

Patients with peroneal nerve injury present with any of the following complaints:

1. Acute foot drop/foot slap
2. Frequent falls
3. Sensory abnormality (numbness/tingling) over lateral half of anterior leg and/or dorsum of the foot

Of note, pain is not a common complaint in these patients and is only present in less than 20 % of this population [2].

History should include inquiry regarding prior trauma, use of external compression device or garment, ankle sprain, knee osteoarthritis, repetitive activity, surgical procedure, or weight loss. Trauma to the fibular head is likely to compromise the peroneal nerve in this area. One epidemiological study of over 211 peripheral neuropathies (in 158 patients) revealed that the peroneal nerve was the most common lower extremity nerve to be injured by a trauma and comprised of 15 % of the traumatic neuropathies [3]. Additionally, treatment of lower leg injury can contribute to peroneal neuropathy. A brace or postsurgical dressing that contacts the area around the fibular neck or over the ankle can compress the nerve.

Forceful inversion, as in rolling the ankle inwards while on the toes, is the typical mechanism of ankle sprain. This could stretch the peroneal nerve leading to injury [4]. Also, the peroneal nerve is stretched with approximately 20 % of knee dislocations [5]. In older patients, a history of osteoarthritis may be significant, as osteophytes may deform the knee joint and result in repetitive traction injury to the nerve [6].

Repetitive squatting, as may occur in a gardener, has been reported to be a cause of fibular neck peroneal neuropathy. Habitually crossing the legs has a similar effect. Prolonged anesthesia can cause a compression or stretching injury to the nerve, depending on the posture of the patient. Prolonged lying in a recumbent position is known to cause injury to the nerve at the fibular head.

Hip injuries and their treatment are potential culprits for an injury to the common peroneal nerve despite their proximal location. This can be due to local compression of the lateral fibers of the sciatic nerve in the posterior hip, which will eventually become the peroneal nerve, or stretch injury to the distal peroneal nerve because it is tethered at the fibular head. The hip injuries implicated in peroneal neuropathy include acetabular fractures, femur fractures, and posterior hip dislocations. These injuries result in peroneal neuropathy in 16–33 % of cases [7].

A careful history should include inquiring about eating disorders or weight loss. Significant or rapid weight loss, such as would occur in anorexia nervosa, bulimia nervosa, and after gastric bypass surgery, is an important risk factor for developing peroneal neuropathy, especially at the fibular neck. This occurs due to a decrease in the protective fat pad around the nerve.

Upon physical examination, one may observe a combination of findings described in the following section. For all peripheral nerve injuries, the more proximal the nerve injury, the broader the scope of the neurological involvement present. Therefore, the following section will review physical examination findings of common peroneal injury from the most proximal to distal locations.

Although peroneal nerve/sciatic nerve injury at the hip is rare, it should still be recognized that injury can occur to sciatic nerve before peroneal nerve divides away from tibial nerve above the knee. It should also be noted that injury to the sciatic nerve is more likely to result in damage to the peroneal fibers than damage to the tibial fibers due to nerve fiber topology. The peroneal fibers have been described to lie more laterally within sciatic nerve sheath when compared with tibial fibers, making it more prone to injury [8].

When the peroneal portion of sciatic nerve is injured, the presentation is very similar to that of peroneal neuropathy at the knee, which will be described below. However, these patients may complain of an additional sensory abnormality over posterolateral portion of the leg due to the involvement of lateral sural cutaneous nerve.

Peroneal neuropathy at the knee affects both superficial and deep peroneal nerves, resulting in weakness or inability to evert (superficial branch) and dorsiflex (deep branch) the foot on manual muscle testing. Sensation should also be tested, and this may reveal abnormality in the distal 2/3 of anterolateral leg and dorsum of the foot including first web space. Deep tendon reflexes should be normal for the patellar and Achilles tendons because the patellar reflex is innervated by the L2–L4 femoral nerve and the Achilles is innervated by the L5–S1.

Gait examination can be extremely valuable in neuropathy. Upon gait observation, one may note foot slap upon heel contact due to inability to activate dorsiflexors and/or an inverted, plantar flexed foot during the swing phase. Complete inability to dorsiflex the foot usually results in a steppage gait to clear the foot drop. Steppage gait is the gait pattern during which one drives the knee higher than usual (as in marching) to clear the plantar flexed foot.

An injury to the L5 nerve root (known as L5 radiculopathy) and an injury to lumbosacral plexus (known as lumbosacral plexopathy) can mimic these signs and symptoms depending on the severity of the injury. To differentiate these two entities from peroneal neuropathy, one can test the strength of posterior tibialis muscle, which is primarily innervated by the L5 nerve root and functions to invert the foot. Since the posterior tibialis is innervated by tibial nerve, one would not expect weakness in inversion if the injury is purely peroneal neuropathy.

Peroneal neuropathy can occur at the ankle with tightly fitted straps or shoes, such as ski boots. Pressure over this area can cause injury to terminal cutaneous branches of deep and/or superficial peroneal nerves. Since it most commonly involves medial and intermediate dorsal cutaneous branches of the superficial peroneal nerve, there will not be detectable weakness on manual testing, and these patients most commonly present with pure sensory deficits. Sensory examination will show numbness over the dorsum of the foot with sparing of first web space (deep peroneal innervation). Gait will be normal in peroneal neuropathy at the ankle.

Peroneal Neuropathy and Gender

A number of cases of common peroneal compression neuropathy have been described in female patients undergoing spontaneous vaginal delivery [9–13]. Most commonly, the mechanism of injury has been proposed to be due to external compression applied by the patient at or around fibular head or due to forceful knee flexion. It is therefore common practice to advise laboring mothers against holding the lateral knees during the spontaneous vaginal labor as this can predispose them to common peroneal nerve injury. However, at least one case has been reported where the mother had developed a foot drop 2 days after delivery despite holding distal the posterior thighs as advised by a gynecologist. The author of this case report concluded that the mechanism of injury was due to a prolonged forceful compression of the posterior thigh proximal to popliteal fossa leading to peroneal neuropathy [13].

Systemic lupus erythematosus has been found to increase one's chance of developing mononeuropathy multiplex with predilection to the peroneal nerve and sural nerves [14]. Since there is approximately a ninefold chance of developing SLE in females [15], as compared to males this phenomenon is noteworthy.

Gastric bypass surgery is increasingly popular among female patients due to the obesity epidemic. According to a report that surveyed gastric bypass surgeries performed in the state of New Hampshire from 1996 to 2007, of the near 3,000 bypasses performed, 78.1 % were for female patients [16]. As in the case of anorexia, sudden weight loss associated with the bypass has been documented to result in an isolated peroneal neuropathy. One case reports bilateral peroneal involvement at the knee severe enough to require a surgical decompression of the nerve [17]. It is also important to recognize the bypass population is more prone to nutritional polyneuropathy than those without the gastric bypass due to resultant vitamin B12 and folate deficiency.

Finally, it should be remembered that women have as much as a five times increased risk of sustaining anterior cruciate ligament (ACL) injury as compared to men [18, 19]. Often, the landing position and postural biomechanics in athletic competition have been attributed as the reason for such gender discrepancy. The female lands more erect due to more anterior center of gravity with less active hamstrings and valgus conformation of the knees, which loads the ACL more than male counterparts. However, peroneal neuropathy is more closely associated with iatrogenic injury from ACL reconstruction procedure than nerve trauma from ACL sprain or tear [20].

Diagnosis and Treatment

When diagnosis is difficult, radiographs at the knee can evaluate for a bone lesion of or near the proximal fibula. An MRI can evaluate for soft tissue changes or a mass lesion as a cause of peroneal neuropathy. If imaging fails to establish a clear diagnosis, then the patient should be referred for electrodiagnostic studies.

Electrodiagnostic testing involves two distinct parts: nerve conduction studies (NCSs) and needle electromyography (EMG). The NCS utilizes electrical conduction to test for severity of injury as well as the location of the injury. It is used to see if the injury involves axons, myelin, or both. Injury involving an axon usually presents with much slower recovery than an injury involving the myelin alone. The second portion is the needle EMG, which involves placing a small pin inside the muscle. This assists in determining the acuity of the problem and further helps detailing the location and severity of the lesion. Combining both nerve conduction and electromyography not only provides information on prognosis for recovery but can help localize the lesion.

Since peroneal neuropathy is a type of peripheral neuropathy, the injury usually recovers spontaneously. Even in the traumatic cases, such as hip dislocation or fracture-dislocation, approximately 60–70 % of sciatic or peroneal injuries will have reasonable spontaneous recovery [21]. Therefore, the treatment is usually prevention of further damage (removal or loosening of tight fitted clothing, braces, or cast, or placement of extra padding over the fibular head) and avoidance of exacerbating activity.

The physician can also assist the patient while healing by ordering an ankle-foot orthoses (AFO) if foot drop and frequent falls are of hindrance to their daily activity. The AFO is available in different materials, and the trim lines and ankle position can be adjusted depending on the severity of the foot drop. These devices range from the newer lightweight carbon fiber energy return AFO to a traditional polypropylene AFO used for more severe foot drop. The device can be solid or articulated at the ankle joint to preserve range of motion.

If pain is a complaint, then topical agents such as capsaicin or lidocaine may be considered before resorting to systemic agents. For neuropathic pain, tricyclic antidepressants, gabapentin, pregabalin, and duloxetine can be considered depending on medical comorbidities and efficacy for the individual patient.

Physical therapy should be offered in conjunction to provide range of motion and strengthening exercises. Pain can be mitigated in physical therapy by application of physical modalities by the therapist such as heat, ice, electrical stimulation, or the use of iontophoresis, which involves application of topical corticosteroid by the physical therapist who then applies electrical current [22]. Observational studies have demonstrated that the density or severity of the foot drop appears to correlate with the prognosis for spontaneous recovery [23]. In the case of significant trauma or nerve transection, referral to neurosurgery will be most effective within the first 5 months post-injury. For nonsurgical specialists, it is important to bear in mind that nerve repair in excess of 5 months is associated with a poorer outcome [24]. Surgical timing depends on the type of peroneal nerve lesion. Sharp lacerations can be repaired immediately, but 2–4 weeks is recommended for blunt lacerations. Nerve trauma due to gunshot wounds is typically surgically explored at 3 months post-injury. In the absence of nerve transection, lesions are usually monitored for 3 months for clinical or electromyographic improvement prior to surgical intervention [25].

Tibial Nerve

Anatomy of the Tibial Nerve

The tibial nerve arises from nerve roots L4–S3 to initially form the sciatic nerve. The sciatic nerve exits from greater sciatic foramen superior to sacrospinous ligament and inferior to piriformis muscle and enters the gluteal area. The sciatic nerve then descends and enters the thigh posteriorly.

The sciatic nerve lies just posterior to adductor magnus muscle and anterior to long head of biceps femoris. As it descends posterior thigh, the tibial portion of sciatic nerve sends motor innervation to all hamstring muscles, except the short head of biceps femoris (from medial: semimembranosus, semitendinosus, and long head of biceps femoris muscles).

The tibial nerve proper arises from the medial branch of the sciatic nerve just proximal to the popliteal fossa. The tibial nerve then descends through the popliteal fossa and enters the posterior compartment of the leg under tendinous arch of soleus muscle. As tibial nerve enters the popliteal fossa, the first cutaneous branch is given off called medial sural cutaneous nerve, which joins with abovementioned lateral sural cutaneous nerve to form sural nerve proper. The tibial nerve in the leg also gives rise to a number of motor branches to supply all posterior leg muscles.

Another cutaneous branch called medial calcaneal nerve branches off the tibial nerve before the tibial nerve enters into the sole of the foot through medial ankle via the flexor retinaculum or the so-called tarsal tunnel. This is the most common site of tibial nerve entrapment and/or neuropathy. Once in the sole of the foot, the tibial nerve bifurcates to the medial plantar nerve and lateral plantar nerves. Both terminal branches are both motor and sensory nerves responsible for intrinsic muscles of the foot as well as sensation of the sole of the foot (medial plantar nerve supplies sensation to medial sole, while lateral plantar nerve is responsible for lateral sole) [1].

Symptoms and Signs

Patients with tibial nerve injury present with any of the following complaints:

1. Sensory abnormality (pain/numbness/tingling) over medial malleolus and/or sole of the foot including the toenails
2. Nocturnal exacerbation of symptoms
3. Motor weakness in plantar flexion and/or inversion of ankle (rare)

Of note, pain is usually the earliest sign of tibial neuropathy, and it may have been already improved at the time of presentation. Currently, there is no universal consensus in terms of which clinical findings make up the diagnosis of tarsal tunnel syndrome (the most common type of tibial neuropathy), although plantar foot dysesthesias and nocturnal exacerbation of symptoms are frequently reported in tarsal tunnel syndrome [26].

History should include inquiry regarding trauma, foot deformities, repetitive motion, and autoimmune disease. Obvious trauma to posterior knee or over the medial malleolus area may cause tibial neuropathy. Foot deformities may be a contributing factor. Some authors have related increased tibial nerve tension and incidences of tarsal tunnel syndrome to pes planus. They explained the tensile force is applied every time the arch collapses upon weight bearing during gait [27]. Injury from repetitive motion occurs most often in the competitive athletic populations that perform repetitive high jumps such as soccer, volleyball, track (high jump), basketball, and gymnastics [28]. For this population, early surgical intervention should be considered. A history of rheumatoid arthritis, ankylosing spondylitis, and Reiter syndrome are known to cause proliferative sinusitis, leading to possible compression of tibial nerve near the tarsal tunnel.

Other than direct trauma to posterior knee, formation of Baker cysts (popliteal cyst) or popliteal artery aneurysms can cause tibial mononeuropathy at this location. In fact, tibial nerve neuropathy at the knee was observed in over one fifth of the Baker cyst cases reported in a comprehensive literature review [29]. Intraneural ganglion cyst has also been documented to cause similar symptoms at the popliteal fossa [30].

On physical examination, one may observe a combination of findings described below. Patients with tibial neuropathy at the knee will likely present with sensory alteration at the heel, sole of the foot, and over the toenails. In addition, such injury may damage the medial sural cutaneous nerve and may cause sensory change in the lateral dorsum of the foot.

When compared to the more distal tarsal tunnel syndrome, these patients will more likely present with weakness in plantar flexion and inversion as motor innervation to posterior leg muscles (gastrocnemius, soleus, and tibialis posterior mainly) is typically involved. In chronic cases, muscle atrophy may be seen as calf atrophy on visual inspection. The Achilles tendon reflex may be diminished or absent on neurological examination. If tibial neuropathy is suspected at the knee, palpation over popliteal fossa may reveal the presence of a Baker cyst.

When tibial nerve injury occurs as the nerve crosses the ankle, it is referred as tarsal tunnel syndrome (TTS). Due to its superficial location, TTS is most commonly caused by trauma sustained during an athletic event to the medial aspect of the ankle joint. As mentioned above, repetitive traction force on the nerve experienced by patients with pes planus can also cause TTS.

Space-occupying lesions, although not related to sports activity, can cause TTS. It is important to recall that tarsal tunnel, or flexor retinaculum of the foot, has very limited space with many structures passing through it [31]. Hypertrophy of flexor hallucis longus tendon has been documented as a cause of TTS. Lipoma, perineural fibrosis, neuroma, presence of an accessory foot muscle, or ganglion cyst has been documented to cause TTS via mass effect [32–34]. Physical examination will likely reveal sensory deficits over sole of the foot and over the toenails, but sparing dorsum and lateral aspect of the foot. Motor weakness is usually difficult to assess as major plantar flexors and inverters of the foot are still intact. In severe cases, one may notice wasting of intrinsic foot muscles. Since motor innervation to posterior leg is intact, the Achilles tendon reflex should be normal. Gait examination should be performed with specific attention paid to ankle or foot deformity.

Tibial Neuropathy and Gender

Any acute or chronic condition that results in inflammation of the surrounding structure of the knee synovial membrane can predispose one to an increased chance of developing a Baker cyst [35, 36]. Rheumatoid arthritis, for which female sex is a risk factor, predisposes the patient to synovitis and may make Baker cyst more likely. Also, as stated above, ACL injury is about five times likely in females. Partial ACL tear often goes undetected due to its asymptomatic nature, but such tears are known to lead one to premature development of OA of the knees [37]. When such OA develops, it can lead to increased risk of OA, leading to theoretical increased risk of compression tibial neuropathy at the knee for females.

Diagnosis and Treatment

In case of trauma-related suspected tibial neuropathy, radiographs of the knee or ankle should be ordered to evaluate for fractures and bone spur formations. Stress fracture can happen to the talus, and this can manifest as medial ankle pain without obvious trauma. When tendinopathy or soft tissue abnormality such as cyst formation or mass compression is suspected, MRI or musculoskeletal ultrasound can be performed to evaluate for such etiology. This may help to identify the presence of accessory muscles or proliferative synovitis.

Differential diagnoses to consider in a patient with suspected TTS include plantar fasciitis and medial ankle sprain. With plantar fasciitis, the patient usually complains of immediate heel pain on the first step out of bed. Medial ankle sprain is usually diagnosed based on history and is extremely rare due to a strong deltoid ligament. Ankle sprain is more likely to cause peroneal neuropathy than tibial neuropathy since it typically affects the lateral ankle.

Electrodiagnostics can be helpful in localizing the lesion. According to one of the most comprehensive literature reviews in Muscle and Nerve, abnormalities of tibial motor NCSs, mixed nerve studies, and/or sensory studies of plantar nerves are possibly effective in diagnosing TTS in TTS-suspected populations based on history and physical (level C, level III evidence) [26]. As for needle electrodiagnostic examination, it is useful to rule out other mimickers of TTS such as S1 radiculopathy or lumbosacral plexopathy.

Treatment is usually conservative and symptomatic for TTS caused by a trauma or stretching injury. Local therapy using topical NSAIDs with concurrent physical therapy can often be helpful to reduce local swelling and ease painful sensation for patients with TTS. Steroid injection to tarsal tunnel can be performed with or without MSK US guidance for symptomatic relief in traumatic cases. Other neuropathic agents such as gabapentin, pregabalin, amitriptyline, and duloxetine can be considered for severely pain in cases of TTS.

If pes planus was noted during the visual inspection or gait examination, traction may contribute to worsening of the neuropathic symptoms. In such cases, an orthotist can determine the appropriate shoe insole to prevent further progression of the disease [38].

When a mass or a cystic lesion is seen on magnetic resonance imaging or musculoskeletal ultrasound at the knee or at the tarsal tunnel, these cases should be most properly addressed by a referral to a surgeon for characterization of the mass and evaluation for possible mass removal or cyst decompression. As for a traumatic tibial nerve injury, the same "5 months to repair" rule applies with electrodiagnostic result suggestive of a transection of the tibial nerve. In the presence of a space-occupying lesion that does not respond to conservative treatment, surgical exploration is warranted.

Baker cysts are usually treated by addressing the underlying etiology. Since a Baker cyst likely follows an inflammatory reaction or meniscal lesion in the knee joint, oral NSAIDs, physical modalities such as icing, compression dressing, or steroid injection can all be considered. These are no longer typically aspirated as they can recur. They can rupture spontaneously, which may result in a temporary increase in calf pain. In severely symptomatic cases, evaluation of and possible surgical repair of an associated meniscal lesion should be considered.

Sural Nerve

Anatomy of the Sural Nerve

The sural nerve, primarily S1 nerve root derived [39], arises by unity of medial sural cutaneous nerve and lateral sural cutaneous nerve (peroneal anastomotic or communicating branch of common fibular nerve). The medial sural cutaneous nerve is a sensory branch of tibial nerve that arises just distal to popliteal fossa, projecting superficially to innervate skin area over lateral head of gastrocnemius muscle.

The lateral sural cutaneous nerve is a sensory branch of common peroneal nerve that arises just before common peroneal nerve wraps around the fibular head, projecting inferomedially to join medial sural cutaneous nerve. Lateral sural cutaneous nerve innervates sensation over posterolateral aspect of proximal leg.

At the distal one third of the posterior leg in between two heads of gastrocnemius, the sural nerve proper forms, and it runs inferiorly in proximity with small saphenous vein while innervating distal one third of posterolateral area of the leg, lateral ankle posterior to lateral malleolus, as well as dorsolateral aspect of the foot and fifth digit [1].

Signs and Symptoms

Patients with sural neuropathy present with a sensory abnormality (pain/numbness/tingling) over:

1. Distal 1/3 posterolateral leg
2. Lateral ankle
3. Dorsolateral aspect of the foot and fifth toe

Of note, the incidence of sural mononeuropathy is rare. Of all nerves, systemic vasculitis is most likely to affect the sural nerve. The sural nerve is purely sensory, and removal results in relatively trivial deficit so this nerve is often employed for location of nerve biopsy or nerve graft harvesting.

History taking should include inquiry regarding a history of autoimmune disorder, connective tissue disorder, trauma, external compression, stretch injury, or varicosities.

Three types of vasculitis are particularly associated with peripheral polyneuropathy known as vasculitic neuropathy [40]. These are (1) ANCA-associated vasculitis (such as Wegener's vasculitis, microscopic polyangiitis, and Churg-Strauss syndrome), (2) polyarteritis nodosa, and (3) mixed cryoglobulinemia. In these vasculitides, the sural nerve has been reported to be the most commonly involved peripheral nerve, making it to the most useful nerve to biopsy for tissue diagnosis.

Although vasculitic polyneuropathy is a common systemic reason for sural neuropathy, the most common cause of isolated sural neuropathy is blunt trauma to lateral ankle and/or foot and its surgical treatment if fracture exists [41]. As a rare cause, blunt trauma to posterior leg has been reported to result in sural mononeuropathy. A case of sural neuropathy has been reported due to a trauma to the gastrocnemius muscle with resultant hematoma formation. This leads to scarification of deep fascia of gastrocnemius muscle and development of sural compression neuropathy [42]. A history of wearing a tight fitted shoes or ankle sprain should be elicited in cases of sural neuropathy [43]. Finally, iatrogenic causes, such as a small saphenous vein stripping procedure for treatment of varicose veins, have been documented to cause sural neuropathy in some case reports [44].

Physical examination may reveal alteration of sensation in posterolateral distal one third of the leg, lateral ankle posterior to lateral malleolus, and dorsolateral foot and fifth toe. Tinel's sign can be performed by tapping anywhere along the nerve, but most commonly tested by tapping area just lateral to musculotendinous junction of the proximal Achilles tendon.

Sural Neuropathy and Gender

Of abovementioned causes of sural nerve neuropathy, the only diagnosis that has known predilection to affect females is mixed cryoglobulinemia (also known as type II or type III cryoglobulinemia). This represents approximately 80 % of all cryoglobulinemia and is usually associated with connective tissue such as systemic lupus and Sjogren syndrome.

Diagnosis and Treatment

In the presence of trauma, radiographs, MRI, and MSK US can all be considered to evaluate the affected area. Electrodiagnostics can be of value, particularly because this enables one to differentiate sural neuropathy from S1 radiculopathy, which can also cause plantar foot dysesthesia. Of note, electrodiagnostics have been found to diagnose about two thirds of cryoglobulinemia.

In the absence of obvious trauma or mass lesions, but with clinical and electrodiagnostic findings of sural neuropathy, vasculitis and connective tissue disorder workup can be considered [45]. This should include serology for ANA (antinuclear antibody), ANCA (antineutrophil cytoplasmic antibodies), cryoglobulins, and RF (rheumatoid factor). Additional serology may include CBC with differential, basic metabolic panel (to assess kidney function), hepatitis C titer (as this may cause cryoglobulinemia), and antibody for double-stranded DNA (specific to SLE) and antibody to nuclear antigens (anti-Sm, Anti-Ro, Anti-La, Anti-RNP). Nerve and/or muscle biopsy can also be done as a confirmatory test when above history, physical, and serology studies are indicative of vasculitic neuropathy. Most frequently, these vasculitides result in axonal injury.

Treatment is usually supportive for pain control with topical or oral NSAIDs and neuropathic pain agents. Of course, modalities such as icing or TENS unit can be considered. For systemic causes of sural neuropathy such as vasculitic neuropathy, treatment should be directed to underlying conditions once diagnosis is made. Referral to the rheumatologist should be considered.

Lumbosacral Plexus

Anatomy of the Lumbosacral Plexus

A group of anterior rami from spinal nerves between L1 and S3 come together to form a bundle of nerves called lumbosacral plexus. This plexus can be anatomically divided to two parts. L1–L4 anterior rami form the upper lumbar plexus, while L5–S3 form the lower lumbosacral plexus, with a small contribution from the L4 nerve roots [1, 46].

The upper lumbar plexus gives rise to the following six nerves:

1. Iliohypogastric nerve (L1: motor to abdominal muscles, sensory to lower anterior abdomen)
2. Ilioinguinal nerve (L1: motor to abdominal muscles, sensation to groin and very small portion of medial thigh as well as upper half of scrotum/labia majora)
3. Genitofemoral nerve (L1–L2: motor to cremasteric muscle, sensation to skin over lower half of scrotum/labia majora as well as anterior-medial thigh)
4. Lateral femoral cutaneous nerve of the thigh (L2–L3, sensation to a large area in anterolateral thigh. It wraps around the anterior superior iliac crest where entrapment or trauma can damage this nerve.)
5. Femoral nerve (L2–L4 posterior division: motor innervation to iliopsoas/hip flexor, pectineus, sartorius, and quadriceps muscles/knee extensors. Sensory innervation to skin area over posterior-medial calf via saphenous nerve)
6. Obturator nerve (L2–L4 anterior division: motor to thigh abductors. Sensory over a small area in medial thigh, inferior to area covered by genitofemoral nerve)

The lower lumbosacral plexus gives rise to the following nerves:

1. Superior gluteal nerve (L4–S1, motor to gluteus medius and minimus muscles/hip abductors)
2. Inferior gluteal nerve (L5–S2, motor to gluteus maximus/hip extensor)
3. Posterior cutaneous nerve of the thigh (S1–S3, sensation over lower buttock and posterior thigh. Exists in proximity to sciatic nerve; therefore, concurrent injury is common).
4. Sciatic nerve (L4–S3, motor to hamstrings and foot/ankle muscles as detailed above. Sensation over the entire leg with exception of posterior-medial leg)

The smaller and less clinically significant nerve to the obturator internus, nerve to the quadratus femoris, and nerve to piriformis will be omitted to focus on the more likely clinical syndromes.

Signs and Symptoms

Signs and symptoms will be better understood when presented in accordance with the anatomic division of the plexus [46].

Patients with upper lumbar plexopathies present any combination of below findings:

1. Weakness in hip flexion (iliopsoas involvement, femoral nerve), knee extension (quad involvement, femoral nerve), and hip abduction (abductor involvement, obturator nerve)
2. Sensory abnormality (pain/numbness/tingling) in anterior pelvic area, anterior, medial, lateral thigh with radiation down to posterior-medial leg (saphenous nerve)
3. Absent or decreased deep tendon reflex at patellar tendon

Patients with lower lumbosacral plexopathies present with any combination of below findings:

1. Weakness in hip extension (gluteus maximus involvement, inferior gluteal nerve), hip abduction (gluteus medius and gluteus minimus involvement, superior gluteal nerve), knee flexion (hamstrings, sciatic nerve), and any motion at the ankle or toes (common, deep, superficial peroneal nerve and tibial nerve)
2. Sensory abnormality (pain/numbness/tingling) in posterior thigh, all of the legs except for saphenous nerve area, and both dorsum and sole of the foot
3. Absent or decreased deep tendon reflex at the Achilles tendon

In lower lumbosacral plexopathies, peroneal nerve fibers are preferentially affected compared to those of the tibial nerve. This is often attributed to fiber topology. Therefore, when patient presents with only foot drop, one should never assume it is an isolated peroneal neuropathy, and a thorough physical exam is critical.

Careful history taking may reveal a history of trauma, coagulopathy, malignancy, radiation, diabetes, infection, and recent childbirth. Direct, penetrating trauma with sharps or gunshots can directly damage the plexus. Also, a secondary injury due to a pelvic or sacral fracture can result in lumbosacral plexopathy due to bony displacement onto the plexus. The incidence and severity of the plexopathy was related to the instability of the fracture (defined as anterior vs. lateral vs. posterior) and the number of anatomic locations fractured [47]. Upper lumbar plexus involvement is more common with direct trauma, while lower lumbosacral plexus involvement is more common with secondary trauma from pelvic fracture due to the proximity to the pelvic bones [48]. Injury near the sacroiliac joint has a much higher risk of lumbosacral plexopathy (upper and lower), when compared with other types of pelvic fracture.

Retroperitoneal hemorrhage or hematoma formation is well known to cause lumbosacral plexopathy and can occur spontaneously in the setting of coagulopathy or anticoagulation therapy [49]. Abdominal and pelvic malignancy has been reported to cause compression plexopathy. Malignancy known to result in the plexopathy can be related to the cancers of the bladder, uterine, prostate, rectum, and pelvic bones. These are usually slowly progressive, and the earliest sign is usually the severe unilateral pelvic pain with or without radiation down to the legs. Often, the pain is described to be constant and achy. It is important to note that pain is rarely described as "burning or sharp." Focal neurological symptoms usually follow the onset of pain.

Since this may be the earliest presenting symptom of the malignancy, it is important to ask about a personal or family history of cancer. About 75 % of cancer-related lumbosacral plexopathy is of local origin and due to direct extension [50]. Colorectal tumor comprises the highest proportion of such causes to account for 20 % of malignancy-induced lumbosacral plexopathy cases.

Breast cancer, an non-intra-abdominal tumor, interestingly accounts for 11 % of lumbosacral plexopathies [50]. Although rarely reported, a recent case series of two patients with Ewing's sarcoma who presented with isolated neuropathic or musculoskeletal pain reminds clinicians to always consider malignancy in the setting of neuropathic pain [51].

Radiation-induced lumbosacral plexopathy typically presents between 1 and 5 years after initial treatment although it has been documented as early as 1 month and as late as 31 years [52]. Unlike malignancy-induced lumbosacral plexopathy, the patient usually presents with painless weakness. Pain is said to be present in only 10 % of patients with radiation-induced lumbosacral plexopathy [53]. Bladder/bowel incontinence is also a common complaint in these patients.

Diabetes is an important cause of lumbosacral plexopathy. Formally known as diabetic amyotrophy, other names used include Bruns-Garland syndrome, proximal diabetic neuropathy, or diabetic lumbosacral radiculoplexopathy [54]. This condition is often characterized by a severe, disabling deep pelvic pain with subsequent development of asymmetric development of weakness in lumbosacral plexus distribution. Exact pathophysiology is still unknown but is suspected to be an immune-mediated inflammatory process [55]. This condition should be recognized as a separate entity from diabetic sensory neuropathy. Only 0.1 % of diabetics have been reported to develop this condition, with a preference for type II diabetics. There has not been any report on a particular discrepancy between the sexes [56]. The majority of cases reported are in patients with 50 years of age or above, and young patients presenting with this condition are very rare [57].

Idiopathic lumbosacral plexopathy or plexitis should be suspected in patients who do not demonstrate an inflammatory, infectious, neoplastic, iatrogenic, or metabolic cause described above. Clinically, these patients present very similarly to diabetic amyotrophy with regard to signs and symptoms but without a history of diabetes. Certain infections including herpes simplex virus, herpes zoster, and Borrelia have been documented to cause lumbosacral plexopathy [58].

Once signs and associated history point to possible lumbosacral plexopathy, physical examination should entail a detailed neurological and musculoskeletal examination documenting any focal neurological deficits or atrophy of the lower limb muscles.

Lumbosacral Plexopathy and Gender

Postpartum maternal lumbosacral plexopathy (intrapartum maternal lumbosacral plexopathy or maternal birth palsy) is a type of compressive plexopathy associated with parturition as the fetal head engages in the pelvic outlet. The plexus can get compressed with head of the fetus, resulting in various symptoms seen in lumbosacral plexopathy. Classically, the lesion is said to occur most commonly at lumbosacral trunk where a branch off of L4 nerve root joins L5 nerve root in this type of injury. Some physicians believe this is due to the fact that the lumbosacral trunk is no longer cushioned by the presence of psoas muscle in between the plexus and head of fetus. These patients may also present with significant weakness in hip abductors due to proximity of the origin of superior gluteal nerve to such compression site.

Although both common peroneal nerve and tibial nerve fibers are compressed within the sciatic nerve sheath in postpartum lumbosacral plexopathy, common peroneal nerve fibers are more commonly injured due to its proximity to the bony structures. As previously discussed earlier in this chapter, careful neurological and musculoskeletal examination can differentiate peroneal neuropathy from a sciatic or plexus lesion. Incidence of maternal obstetric injury is reported to be 1:2,000 to 1:6,400 deliveries [59].

Due to the proximity of the structures to the plexus, both benign and malignant gynecological cancers are known to cause compressive plexopathy. Leiomyoma, a benign tumor of the uterus, has been documented as a cause of such gynecological neoplastic plexopathy although rare [60]. The majority of such plexopathy is still due to malignancy of the intra-abdominal cavity, retroperitoneal space, and pelvis including malignancy of the ovary, cervix, and uterus. It is very important to remember breast cancer is a significant cause of bilateral lumbosacral plexopathy, as 25 % of bilateral lumbosacral plexopathy is due to breast cancer metastasis [61].

Finally, it should be remembered the risk of lumbosacral plexopathy is increased after a kidney transplant in hypertensive females due to unknown mechanisms. The kidney transplant is proposed to result in the plexopathy either due to uremic nerve injury or direct surgical injury. Female patients who undergo a kidney transplant have been reported to have 1.32 times increased risk when compared to male counterparts [62].

Diagnosis and Treatment

After careful history and physical examination, one should consider imaging studies with radiographs, CT, MRI, or ultrasound. Referral to a physiatrist or a neurologist for electrodiagnostic testing is helpful. The nerve conduction portion will clarify whether the injury is due to damage to a nerve root, plexus, or peripheral nerves. The electromyography portion will not only help determine the temporal course of the pathology, but EMG could potentially identify a myokymic discharge pattern in an affected weak muscle. This is pathognomonic for radiation-induced lumbosacral plexopathy or nerve injury. Imaging will likely be necessary to rule out a neoplastic cause. If malignancy is suspected, serology for tumor markers along with basic labs including CBC, CMP, and ESR should be performed.

Treatment of lumbosacral plexopathy usually requires referral to an appropriate specialist once diagnosis is made. Since most trauma-induced or bleeding-related lumbosacral plexopathy presents to the emergency room, these patients should be promptly evaluated by one of the surgical specialists. If workup reveals malignancy, referral to an oncologist will determine treatment options.

When history and physical are strongly suggestive of diabetic amyotrophy, it is important to remember that tight glycemic control is associated with favorable outcomes. Prognosis is usually good with 60 % of patients improving over 1–2 years time. However, some of these patients may still have some residual weakness or sensory alteration afterwards. Relapse has been documented in as many as 20 % of these patients [63]. IVIG has been tried with some success, suggesting a possible inflammatory mechanism of diabetic amyotrophy [64, 65].

In idiopathic lumbosacral plexopathy, it is important to recognize that this is a diagnosis of exclusion and clinical diagnostic criteria are as follows:

1. Acute onset of leg pain followed by weakness and/or atrophy (similar to diabetic amyotrophy)
2. EMG revealing a patchy pattern of denervation in the distribution of part or all of the lumbosacral plexus but not in the paraspinal muscles (However, diabetic amyotrophy may involve paraspinals due to possible involvement of nerve roots.)
3. Exclusion of all other causes of lumbosacral plexopathy discussed above
4. Eventual recovery, often incomplete, over the following months to years [57] similar to diabetic amyotrophy

When some of the above-listed life-threatening causes of lumbosacral plexus injury have been ruled out, the practitioner should consider supportive use of orthosis or bracing to improve functional independence in these patients with motor weakness or abnormality in proprioception. Physical therapy should be considered to help facilitate recovery of the strength lost and to teach compensatory techniques.

Conclusion

Peroneal neuropathy is by far the most common lower limb entrapment neuropathy, likely due to its vulnerability at the fibular head. Care should be taken during parturition to prevent compromise of this nerve or any of the other lower limb nerves. Treatment is largely symptomatic. Physical therapy is of paramount importance, but orthotics may be helpful in the case of prolonged impairment.

References

1. Jenkins DB, Hollinshead WH. Hollinshead's functional anatomy of the limbs and back. The Leg. 8th ed. Philadelphia: WB Saunders; 2002.
2. Katirji MB, Wilbourn AJ. Common peroneal mononeuropathy: a clinical and electrophysiologic study of 116 lesions. Neurology. 1988;38(11):1723–8.

3. Ciaramitaro P, Mondelli M, Logullo F, et al. Traumatic peripheral nerve injuries: epidemiological findings, neuropathic pain and quality of life in 158 patients. J Peripher Nerv Syst. 2010;15(2):120–7.
4. Strazar R, White CP, Bain J. Foot reanimation via nerve transfer to the peroneal nerve using the nerve branch to the lateral gastrocnemius: case report. J Plast Reconstr Aesthet Surg. 2011; 64(10):1380–2.
5. Feliciano DVME, Mattox KL. Injury to the peripheral nerves. 3rd ed. Appleton & Lange: Stamford; 1996.
6. Fetzer GB, Prather H, Gelberman RH, Clohisy JC. Progressive peroneal nerve palsy in a varus arthritic knee. A case report. J Bone Joint Surg Am. 2004;86-A(7):1538–40.
7. Epps Jr CH. General principles in the prevention of surgical complications in orthopedic surgery. Clin Orthop Relat Res. 1987;222:5–20.
8. Feinberg JH, Nadler SF, Krivickas LS. Peripheral nerve injuries in the athlete. Sports Med. 1997;24(6):385–408.
9. Colachis 3rd SC, Pease WS, Johnson EW. A preventable cause of foot drop during childbirth. Am J Obstet Gynecol. 1994;171(1):270–2.
10. Adornato BT, Carlini WG. "Pushing palsy": a case of self-induced bilateral peroneal palsy during natural childbirth. Neurology. 1992;42(4):936–7.
11. Sahai-Srivastava S, Amezcua L. Compressive neuropathies complicating normal childbirth: case report and literature review. Birth. 2007;34(2):173–5.
12. Hashim SS, Adekanmi O. Bilateral foot drop following a normal vaginal delivery in a birthing pool. J Obstet Gynaecol. 2007;27(6):623–4.
13. Radawski MM, Strakowski JA, Johnson EW. Acute common peroneal neuropathy due to hand positioning in normal labor and delivery. Obstet Gynecol. 2011;118(2 Pt 2):421–3.
14. Florica B, Aghdassi E, Su J, Gladman DD, Urowitz MB, Fortin PR. Peripheral neuropathy in patients with systemic lupus erythematosus. Semin Arthritis Rheum. 2011;41(2):203–11.
15. Danchenko N, Satia JA, Anthony MS. Epidemiology of systemic lupus erythematosus: a comparison of worldwide disease burden. Lupus. 2006;15(5):308–18.
16. Cherala SS. Gastric bypass surgeries in New Hampshire, 1996-2007. Prev Chronic Dis. 2012; 9:E72.
17. Elias WJ, Pouratian N, Oskouian RJ, Schirmer B, Burns T. Peroneal neuropathy following successful bariatric surgery. Case report and review of the literature. J Neurosurg. 2006;105(4): 631–5.
18. Walden M, Hagglund M, Werner J, Ekstrand J. The epidemiology of anterior cruciate ligament injury in football (soccer): a review of the literature from a gender-related perspective. Knee Surg Sports Traumatol Arthrosc. 2011;19(1):3–10.
19. Agel J, Arendt EA, Bershadsky B. Anterior cruciate ligament injury in national collegiate athletic association basketball and soccer: a 13-year review. Am J Sports Med. 2005;33(4): 524–30.
20. Hall MP, Ryzewicz M, Walsh PJ, Sherman OH. Risk of iatrogenic injury to the peroneal nerve during posterolateral femoral tunnel placement in double-bundle anterior cruciate ligament reconstruction. Am J Sports Med. 2009;37(1):109–13.
21. Cornwall R, Radomisli TE. Nerve injury in traumatic dislocation of the hip. Clin Orthop Relat Res. 2000;377:84–91.
22. Edwards Jr PH, Wright ML, Hartman JF. A practical approach for the differential diagnosis of chronic leg pain in the athlete. Am J Sports Med. 2005;33(8):1241–9.
23. Pigott TJ, Jefferson D. Idiopathic common peroneal nerve palsy–a review of thirteen cases. Br J Neurosurg. 1991;5(1):7–11.
24. Bonnard C, Anastakis DJ, van Melle G, Narakas AO. Isolated and combined lesions of the axillary nerve. A review of 146 cases. J Bone Joint Surg Br. 1999;81(2):212–7.
25. Kim DH, Murovic JA, Tiel RL, Kline DJ. Management and outcomes in 318 operative common peroneal nerve lesions at the Louisiana State University Health Sciences Center. Neurosurgery. 2004;54(6):1421–8.

26. Patel AT, Gaines K, Malamut R, Park TA, Toro DR, Holland N. Usefulness of electrodiagnostic techniques in the evaluation of suspected tarsal tunnel syndrome: an evidence-based review. Muscle Nerve. 2005;32(2):236–40.
27. Daniels TR, Lau JT, Hearn TC. The effects of foot position and load on tibial nerve tension. Foot Ankle Int. 1998;19(2):73–8.
28. Kinoshita M, Okuda R, Yasuda T, Abe M. Tarsal tunnel syndrome in athletes. Am J Sports Med. 2006;34(8):1307–12.
29. Sanchez JE, Conkling N, Labropoulos N. Compression syndromes of the popliteal neurovascular bundle due to Baker cyst. J Vasc Surg. 2011;54(6):1821–9.
30. Adn M, Hamlat A, Morandi X, Guegan Y. Intraneural ganglion cyst of the tibial nerve. Acta Neurochir. 2006;148(8):885–9. discussion 889-890.
31. Rodriguez D, Devos Bevernage B, Maldague P, Deleu PA, Leemrijse T. Tarsal tunnel syndrome and flexor hallucis longus tendon hypertrophy. Orthop Traumatol Surg Res. 2010;96(7): 829–31.
32. Spinner RJ. Re: Tarsal tunnel syndrome caused by epineural ganglion of the posterior tibial nerve. J Foot Ankle Surg. 2005;44(1):74–5.
33. Kinoshita M, Okuda R, Morikawa J, Abe M. Tarsal tunnel syndrome associated with an accessory muscle. Foot Ankle Int. 2003;24(2):132–6.
34. Chen WS. Lipoma responsible for tarsal tunnel syndrome. Apropos of 2 cases. Rev Chir Orthop Reparatrice Appar Mot. 1992;78(4):251–4.
35. Sansone V, de Ponti A, Paluello GM, del Maschio A. Popliteal cysts and associated disorders of the knee. Critical review with MR imaging. Int Orthop. 1995;19(5):275–9.
36. Labropoulos N, Shifrin DA, Paxinos O. New insights into the development of popliteal cysts. Br J Surg. 2004;91(10):1313–8.
37. Lohmander LS, Ostenberg A, Englund M, Roos H. High prevalence of knee osteoarthritis, pain, and functional limitations in female soccer players twelve years after anterior cruciate ligament injury. Arthritis Rheum. 2004;50(10):3145–52.
38. Hudes K. Conservative management of a case of medial epicondylosis in a recreational squash player. J Can Chiropr Assoc. 2011;55(1):26–31.
39. Liguori R, Krarup C, Trojaborg W. Determination of the segmental sensory and motor innervation of the lumbosacral spinal nerves. An electrophysiological study. Brain. 1992;115(Pt 3): 915–34.
40. Davies L, Spies JM, Pollard JD, McLeod JG. Vasculitis confined to peripheral nerves. Brain. 1996;119(Pt 5):1441–8.
41. Stickler DE, Morley KN, Massey EW. Sural neuropathy: etiologies and predisposing factors. Muscle Nerve. 2006;34(4):482–4.
42. Bryan 3rd BM, Lutz GE, O'Brien SJ. Sural nerve entrapment after injury to the gastrocnemius: a case report. Arch Phys Med Rehabil. 1999;80(5):604–6.
43. Schuchmann JA. Isolated sural neuropathy: report of two cases. Arch Phys Med Rehabil. 1980; 61(7):329–31.
44. Simonetti S, Bianchi S, Martinoli C. Neurophysiological and ultrasound findings in sural nerve lesions following stripping of the small saphenous vein. Muscle Nerve. 1999;22(12):1724–6.
45. Solomon DH, Kavanaugh AJ, Schur PH. Evidence-based guidelines for the use of immunologic tests: antinuclear antibody testing. Arthritis Rheum. 2002;47(4):434–44.
46. Preston DC, Shapiro BE. Electromyography and neuromuscular disorders: clinical-electrophysiologic correlations. 2nd ed. Philadelphia: Elsevier Butterworth Heinemann; 2005.
47. Jang DH, Byun SH, Jeon JY, Lee SJ. The relationship between lumbosacral plexopathy and pelvic fractures. Am J Phys Med Rehabil. 2011;90(9):707–12.
48. Chiou-Tan FY, Kemp Jr K, Elfenbaum M, Chan KT, Song J. Lumbosacral plexopathy in gunshot wounds and motor vehicle accidents: comparison of electrophysiologic findings. Am J Phys Med Rehabil. 2001;80(4):280–5. quiz 286-288.
49. Ozcakar L, Sivri A, Aydinli M, Tavil Y. Lumbosacral plexopathy as the harbinger of a silent retroperitoneal hematoma. South Med J. 2003;96(1):109–10.

50. Evans BA, Stevens JC, Dyck PJ. Lumbosacral plexus neuropathy. Neurology. 1981;31(10):1327–30.
51. Esser SM, Baima J. Ewing sarcoma causing back and leg pain in 2 patients. PM R. 2012;4(4):317–21.
52. Thomas JE, Cascino TL, Earle JD. Differential diagnosis between radiation and tumor plexopathy of the pelvis. Neurology. 1985;35(1):1–7.
53. Abu-Rustum NR, Rajbhandari D, Glusman S, Massad LS. Acute lower extremity paralysis following radiation therapy for cervical cancer. Gynecol Oncol. 1999;75(1):152–4.
54. Garland H. Diabetic amyotrophy. Br Med J. 1955;2(4951):1287–90.
55. Kawamura N, Dyck PJ, Schmeichel AM, Engelstad JK, Low PA, Dyck PJ. Inflammatory mediators in diabetic and non-diabetic lumbosacral radiculoplexus neuropathy. Acta Neuropathol. 2008;115(2):231–9.
56. Bastron JA, Thomas JE. Diabetic polyradiculopathy: clinical and electromyographic findings in 105 patients. Mayo Clin Proc. 1981;56(12):725–32.
57. Casey EB, Harrison MJ. Diabetic amyotrophy: a follow-up study. Br Med J. 1972;1(5801):656–9.
58. Yee T. Recurrent idiopathic lumbosacral plexopathy. Muscle Nerve. 2000;23(9):1439–42.
59. Hill EC. Maternal obstetric paralysis. Am J Obstet Gynecol. 1962;83:1452–60.
60. Felice KJ, Donaldson JO. Lumbosacral plexopathy due to benign uterine leiomyoma. Neurology. 1995;45(10):1943–4.
61. Chad DA, Recht LD. Neuromuscular complications of systemic cancer. Neurol Clin. 1991;9(4):901–18.
62. Nikoobakht M, Mahboobi A, Saraji A, et al. Pelvic nerve neuropathy after kidney transplantation. Transplant Proc. 2007;39(4):1108–10.
63. Coppack SW, Watkins PJ. The natural history of diabetic femoral neuropathy. Q J Med. 1991;79(288):307–13.
64. Krendel DA, Costigan DA, Hopkins LC. Successful treatment of neuropathies in patients with diabetes mellitus. Arch Neurol. 1995;52(11):1053–61.
65. Williams CP, Child DF, Hudson PR, et al. Inappropriate phosphate excretion in idiopathic hypercalciuria: the key to a common cause and future treatment? J Clin Pathol. 1996;49(11):881–8.
66. Kim S, Choi J-Y, Huh Y-M, Song H-T, Lee S-A, Kim SM, Suh J-S. Role of magnetic resonance imaging in entrapment and compressive neuropathy—what, where, and how to see the peripheral nerves on the musculoskeletal magnetic resonance image: part 1. Overview and lower extremity. Eur Radiol. 2007;17(1):139–49.

Chapter 6
Osteoarthritis and Gender-Specific Joint Replacement

Dean Ehrlich, Nicholas Colacchio, and Eric L. Smith

Abstract Osteoarthritis is a debilitating joint disease which primarily affects women for reasons that remain unclear. There are many treatment options available to address the pain and loss of function in osteoarthritis, ranging from noninvasive physical therapy to total joint replacement. Most of the treatments are gender neutral, but recently knee implants have been marketed to women as gender specific. This chapter reviews the most recent literature on these topics.

The literature suggests that gender-neutral knee and hip implants used for total joint replacements are equally beneficial in both men and women. Gender-specific knee implants have not shown any increased benefit in short-term studies, and it remains to be seen how they will compare to gender-neutral knee implants in the long term. There are no gender-specific hip implants on the market, and there is not a clear consensus about whether the production of a gender-specific hip is necessary.

Keywords Osteoarthritis • Joint replacement • Total knee replacement • Total knee arthroplasty • Total hip replacement • Total hip arthroplasty • Gender • Women

Abbreviations

AAOS	American Academy of Orthopaedic Surgeons
ACR	American College of Rheumatology
HRT	Hormone Replacement Therapy

D. Ehrlich, BS
Department of Orthopaedics, Tufts University School of Medicine,
Tufts Medical Center, Boston, MA 02111, USA
e-mail: dean.ehrlich@tufts.edu

N. Colacchio, MD • E.L. Smith, MD (✉)
Department of Orthopaedics, Tufts Medical Center,
800 Washington Street, 306, Boston, MA 02111, USA
e-mail: ncolaccchio@tuftsmedicalcenter.org; esmith@tuftsmedicalcenter.org

E. Mody, E. Matzkin (eds.), *Musculoskeletal Health in Women*,
DOI 10.1007/978-1-4471-4712-1_6, © Springer-Verlag London 2014

NSAIDs	Nonsteroidal Anti-Inflammatory Drugs
OA	Osteoarthritis
OP	Osteoporosis
THR	Total Hip Replacement
TKR	Total Knee Replacement

Introduction

Osteoarthritis (OA) is a degenerative joint disease caused by a combination of genetic, mechanical, and inflammatory factors that are not well understood. Most patients present with cartilage destruction, narrowed joint space, and osteophyte formation, which result in pain and loss of function (Fig. 6.1a, b) [1]. OA is the most prevalent joint disorder [2] and leading cause of disability in the USA [3]. The cost of treatment and loss of function makes OA a huge financial burden to individuals and society [4].

Epidemiologic studies consistently show that women have an increased risk over men for developing knee and hip OA, the two most common forms of the disease [5, 6]. To make matters worse, women with OA generally express higher levels of pain than men with OA, even when compared with men who have the same radiographic severity of OA [7–9]. The pain of OA not only limits physical function but has psychological impact as well; women with OA report lower satisfaction in life than women without OA [10].

With such widespread impact, it is important for women to understand current risk factors and prevention for development of OA, updated recommendations for management of OA, and whether women should be receiving different or supplementary treatment to achieve the most optimum outcomes.

In this chapter, we will explore:

- Some of the potential reasons why women are at greater risk for developing OA, with specific attention paid to knee OA and hip OA
- Treatments prior to total joint replacement for knee and hip OA
  - Treatments specific to women
- Total joint replacement as an option for treating knee and hip OA
  - Do women and men have similar outcomes with standard total joint replacement?
  - Are gender-specific replacements necessary?

Women and Osteoarthritis

Loss of cartilage is an important risk factor for developing osteoarthritis (OA) [1]. Women are at an especially high risk because they generally lose cartilage in the knee at a faster rate than men [11]. The influence of estrogen on cartilage loss and OA development as estrogen levels change during menopause has been investigated

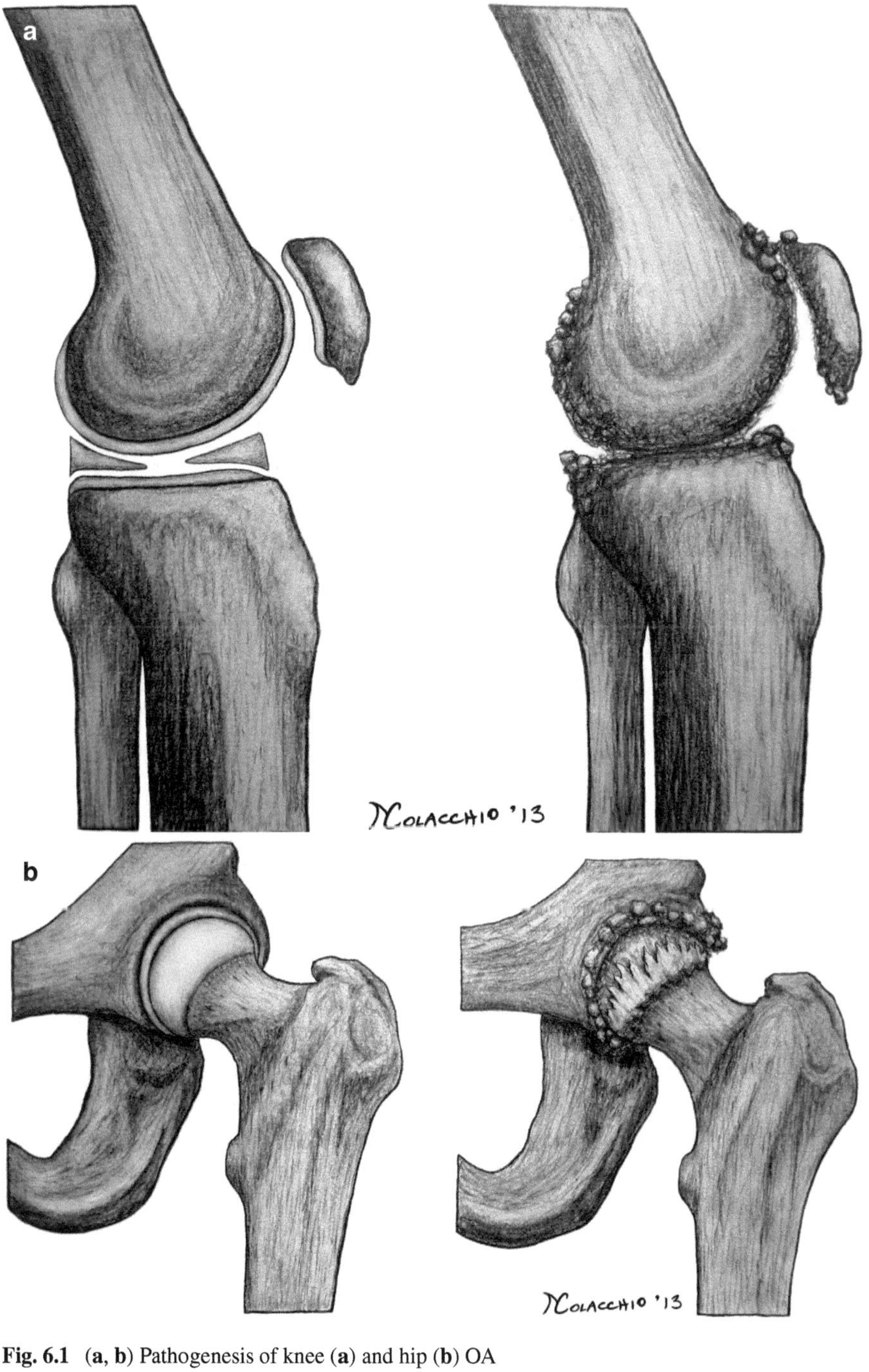

Fig. 6.1 (**a**, **b**) Pathogenesis of knee (**a**) and hip (**b**) OA

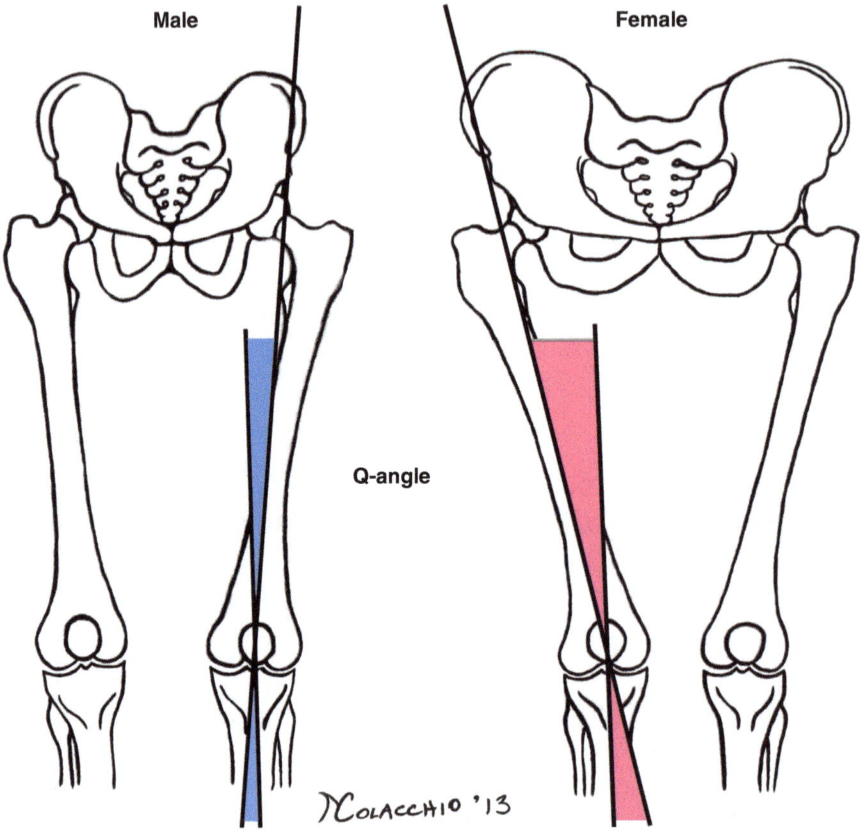

Fig. 6.2 Q angle. Draw a line from the anterior superior iliac spine of the hip to the patella, then another line from the patella to the tibial tuberosity. The Q angle is measured in between these *two lines*

[12]. Despite two decades of research, the impact of estrogen levels on OA is inconclusive and the mechanism by which estrogen physiologically affects cartilage remains elusive [13]. Interestingly, genetic variations in genes for estrogen receptors have been associated with either higher or lower rates of OA, implying that the estrogen hormone does play some role in OA [14, 15]. Much of the research on estrogen and OA has focused on hormone replacement therapy (HRT), which will be discussed in detail in the treatment section; however, results of HRT on OA are similarly equivocal.

There are many anatomic differences between men and women in the knee and hip joints. Joint malalignment has been shown to negatively affect the progression of OA [16, 17], and therefore, different anatomic factors in women could potentially predispose women to higher levels of OA than men. Women generally have wider hips [18] and a larger Q angle than men (Fig. 6.2) [19, 20]. Women also have a thinner patella [21], and differences in the development of knee cartilage from an early age have been noted, which could account for the decreased knee cartilage thickness that is seen in adult women [22]. Women are predisposed to a higher rate of anterior cruciate ligament injuries [23, 24], which has been associated with knee OA later in life [25, 26].

Obesity is a risk factor for development of OA for men and women [27], impacting joints mechanically and hormonally. The knee absorbs between two and five times the normal body weight of an individual, so the increased body weight in obesity is hypothesized to add significant mechanical pressure to the knee with each step taken [27]. However, the increased mechanical strain can only explain part of the increase of OA with obesity, because there is also an association between obesity and increased risk of hand OA [28–30]. Interestingly, obesity is not associated with an increase in hip OA [30, 31]. One possible hormonal explanation for the correlation between obesity and OA is that the increased adipose (fat) tissue releases certain chemical signals, which could systemically affect the joints of the body. One chemical hypothesized to be involved is called leptin, which is released by adipose cells [28]. Women generally have a higher percentage of body fat than men [32], so this could be part of the explanation for why women have a higher prevalence of OA. In fact, one recent study found that obese women having higher leptin levels were associated with an increased chance of developing knee OA [33]. Indeed, weight loss is a recommended treatment for OA for both men and women (see treatment section).

The prevalence of osteoporosis (OP) (disease of decreased bone density) is much higher in women [34], and there have been links between OP and OA, although the exact relationship is uncertain. According to a recent report by the National Institutes of Health Osteoporosis and Related Bone Diseases National Resource Center [35], patients with OA may be less likely to develop OP. However, other studies have found contradictory results, arguing that OP is not looked for often enough in patients with OA [36, 37]. The exact relationship remains undefined at this point, and future research will help us determine whether OA and OP are risk factors for each other and how best to optimize prevention and treatment for these two similar but very different disease processes.

Treatments Prior to Total Joint Replacement for Knee and Hip OA

Osteoarthritis (OA) is a progressive, degenerative disease with a wide range of treatment options for patients at different stages of disease ranging from non-pharmacologic methods to total joint replacement. This section will cover treatments that are generally used before resorting to knee or hip implant. These non-replacement treatments for knee and hip OA are generally the same and will be presented as such except where noted. The core treatments for knee and hip OA do not vary between men and women, but some alternative therapeutic methods studied in women will be presented as well at the end of this section.

The American Academy of Orthopaedic Surgeons (AAOS) has put out a set of guidelines for treatment of knee OA [38]. In its guidelines, the AAOS recommends, suggests, provides the option, or remains inconclusive—for or against treatments—based on the level of evidence in the literature and based on the balance of benefit versus harm for a particular treatment.

The non-pharmacologic therapeutic methods recommended by the AAOS are participation in self-management programs, strength training, low-impact aerobic fitness, neuromuscular education, and physical activity in accordance with national guidelines. The AAOS suggests weight loss for patients with symptomatic OA and a body mass index ≥25. The American College of Rheumatology (ACR) additionally suggests psychosocial intervention, Tai Chi, walking aids as needed, and thermal agents plus manual therapy with exercise supervised by a physical therapist [39]. The AAOS, however, found inconclusive evidence on the use of manual therapy. There was also inconclusive evidence on the use of valgus force-directing knee braces and physical agents like nerve stimulation or electromagnetic therapy. The AAOS recommends *against* the use of glucosamine and chondroitin as well as the use of acupuncture. The AAOS suggests *against* using a lateral wedge insole for symptomatic medial compartment knee OA.

In terms of pharmacological interventions, the AAOS recommends Non-Steroidal Anti-Inflammatory Drugs (NSAIDs) or Tramadol for pain relief. This recommendation includes both non-selective NSAIDs and selective NSAIDS (cyclo-oxygenase-2 inhibitors). The evidence was inconclusive on the use of acetaminophen (new FDA maximum of 3 grams/day), opioids, or pain patches, based on a lack of relevant studies in the literature. The ACR, however, strongly recommends the use of opioid analgesics for those patients with pain refractory to standard pharmacological treatments and who are not willing or able to undergo total joint replacement [39].

The AAOS guidelines also review procedural treatments for knee OA that are less invasive than surgery. Based on a lack of evidence, the guidelines are inconclusive on the use of corticosteroid intra-articular injection, growth factor injection, or platelet rich plasma injection. There is a strong recommendation *against* the use of hyaluronic acid (viscosupplementation) intra-articular injection based on lack of clear evidence showing benefit. The AAOS also suggests *against* the use of needle lavage based on lack of benefit to patients.

There are also a number of surgical approaches, prior to total replacement, that can be used to treat knee OA, often in patients with specific conditions. For patients with medial compartment knee OA, the AAOS gives the option for a valgus producing proximal tibial osteotomy based on limited evidence. For patients with knee OA and a torn meniscus, the AAOS remains inconclusive on arthroscopic partial meniscectomy. In patients with a primary diagnosis of knee OA, the AAOS makes a strong recommendation *against* arthroscopy with lavage and/or debridement based on lacking beneficial evidence and risks from surgery. Also, despite a lack of reliable evidence in the literature, the AAOS workgroup came to a consensus recommendation based on expert opinion that the use of a free-floating interpositional device in patients with medial compartment knee OA is *not* recommended.

There are some treatments for OA specific to women as well, but these are not the primary treatments used in general for knee or hip OA. The most well-known treatment that has been used in the past but has now fallen out of favor is hormone replacement therapy (HRT). There may be a slight reduction in risk of OA [40] with the use of HRT, but the risks of cancer, cardiovascular disease, venous thromboembolism, and gallbladder disease, among other conditions, significantly outweigh the benefits [41]. And some studies on HRT have also shown no benefit of HRT on OA

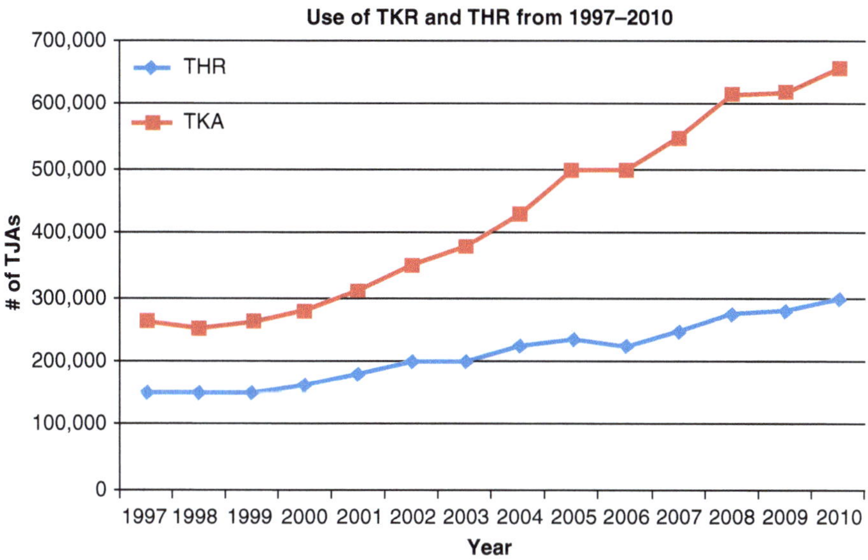

Fig. 6.3 The number of total knee and total hip replacements per year from 1997 to 2010 (Source: Healthcare cost and utilization project (HCUP), Nationwide inpatient sample (NIS) [4])

or have even suggested a deleterious effect of HRT on OA [40, 42]. For young athletic women, treatment with topical NSAIDs as a first-line treatment has been suggested to avoid the gastrointestinal and cardiovascular risks of oral NSAIDs [43]. Finally, there has been research indicating a beneficial effect on overall knee OA outcome with the incorporation of balancing exercises as a compliment to a standard strength-training regime [44].

Total Joint Replacement

Introduction

Total joint replacement is a last resort for patients who have failed nonoperative treatments for osteoarthritis (OA). The total joint replacement procedure has become commonplace in the USA; over 900,000 total knee and hip arthroplasties were performed last year [4], a number that is predicted to rise to 3.8 million in the year 2030 (Fig. 6.3) [45]. The vast majority of patients receive marked functional improvement, and the rate of feared complication is remarkably low [46–48]. Total joint replacement is generally performed on middle-aged to elderly patients, with about 90 % of procedures being done in people aged 45–84 [4].

The standard procedure for total joint replacement is a relatively simple concept. For the knee, shave the arthritic areas of the distal femur (thigh) bone and tibia (shin) bone and replace them with metal, ceramic, or plastic implants (Fig. 6.4a, b).

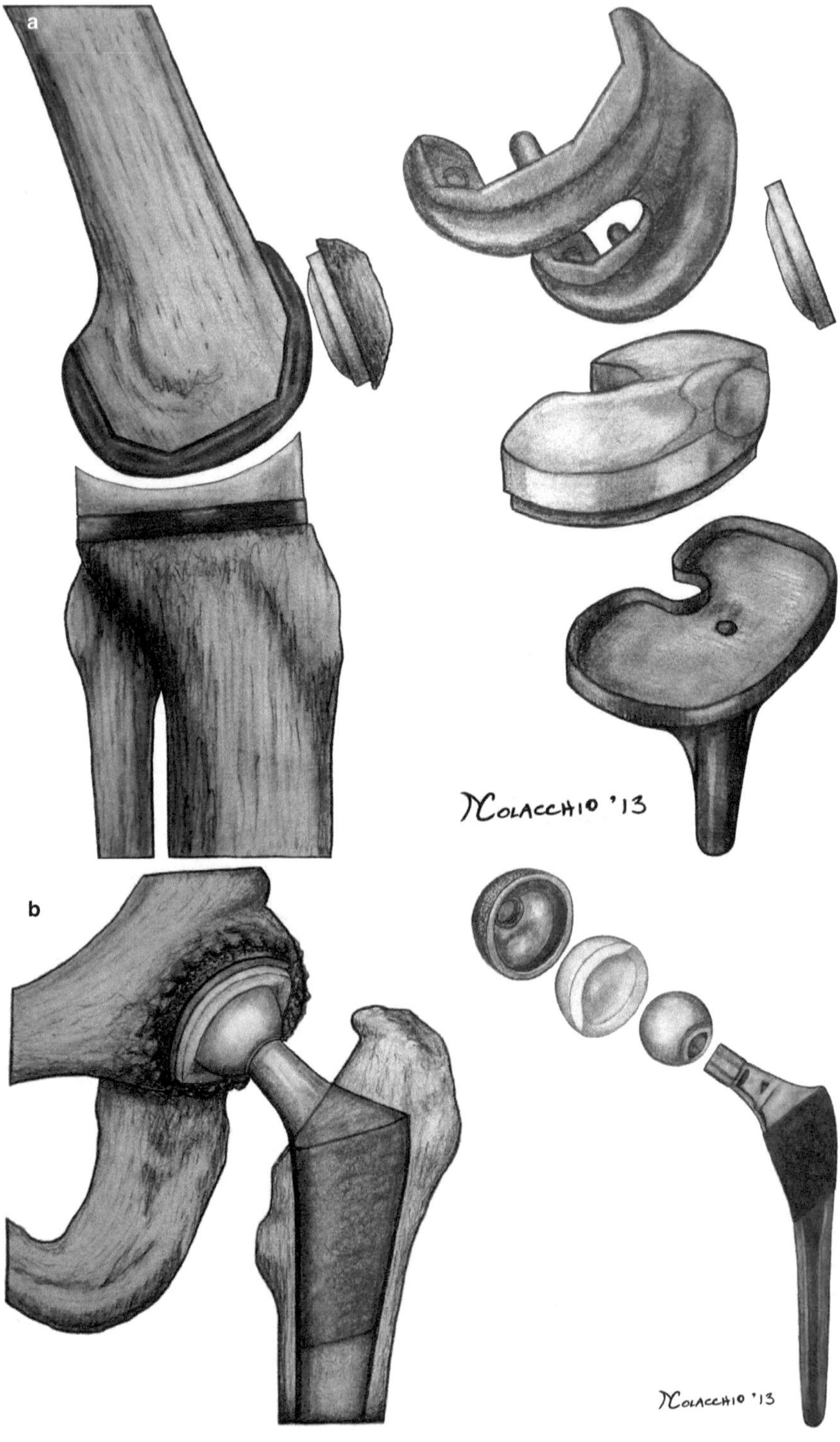

Fig. 6.4 (**a**, **b**) Total knee replacement (**a**). Total hip replacement (**b**)

For the hip, shave the arthritic areas of the proximal femur bone and the acetabulum (hip socket) and replace them with metal, ceramic, or plastic parts. Two of the most important factors for successful joint replacement are alignment and fit of the implanted parts. Certainly much of the success is attributed to the skill of the orthopedic surgeon, but it is also imperative that the proper size and make of the implant fit well with the natural anatomy of the joint of the specific patient. There are many companies making implants of various sizes to fit people with knees and hips of different dimensions, but these implants are generally designed based on average knee and hip dimensions without regard to gender differences.

Evidence-based studies have shown that there are anatomic differences between the male and female knee and hip joints, which could impact total joint replacement. Women have a wider pelvis [18], more bowing of the femoral shaft [49], and a larger Q angle than men (Fig. 6.2) [19, 20]. Within the knee joint, women generally have smaller femoral and tibial condylar heights, narrower transepicondylar widths, a narrower femur, and smaller patella [50]. The rotation of the femur on the tibia is also slightly different in the female than the male knee [51]. Within the hip joint, women generally have a smaller acetabulum, a shorter femoral head, and increased anteversion (femoral neck leans forward causing internal rotation of the knee and foot) [52]. Whether these anatomic differences lead to different outcomes with a generic knee or hip implant or whether they warrant gender-specific knee or hip implants is the subject of the upcoming sections.

For over 6 years, implant companies have been manufacturing knee implants specifically designed for the female anatomy. Unlike the pharmaceutical industry, in which medications must go through a long process before approval by the Food and Drug Administration, small changes in implant design can be brought to the market sooner. With the proven anatomic differences between men and women, implant companies are making and marketing more expensive women-specific implants, which may or may not lead to better outcomes. The women-specific knee implants are generally smaller, narrower, and have a deeper trochlear groove than their generic counterparts, to match the female anatomy.

Knee

Introduction

Osteoarthritis (OA) pain refractory to nonoperative treatments is an indication for total knee replacement [53], and women have higher rates of OA than men [5, 6]. Women undergo more total knee replacements than men [4]. However, it has been shown that the proportion of women who need a knee replacement and actually get one is significantly lower than the proportion of men who need a knee replacement and receive one [54, 55]. Women also generally have worse pain, poorer function [56], and worse quadriceps (front thigh) muscle strength [55] prior to knee replacement. According to the literature, highest postoperative success after TKR can be best predicted by better preoperative knee function scores and quadriceps muscle strength [46]. Therefore, it is important for both the doctor and female patient to recognize that women generally wait longer to have a TKR than men and that it may be advantageous to undergo TKR earlier in the disease process.

Outcomes with Generic Total Knee Replacement (TKR)

While the male and female knee anatomy does have differences, are they clinically significant when comparing outcomes after total knee replacement with *generic* knee implants? To summarize a growing body of evidence-based, peer-reviewed literature, the answer to this question is no; there is no significant difference between the outcomes in women and men with generic knee implants. Women achieve similar functional improvement in a range of different physical tests, equal pain and flexion improvement, and in some cases achieve greater improvement than men with generic knee implants [50, 57–61]. Although there was one study that showed poorer patellofemoral function in women versus men with standard TKR [62] and another study suggesting that African-American women show poorer recovery than other groups [63], most studies report no significant functional difference and outcome, and a recent study showed that women recover faster than men after generic TKR [64]. Overall, the literature suggests that gender does not impact clinical outcome after standard TKR.

Outcomes with Gender-Specific TKR

Despite no significant difference in the outcome after TKR between men and women with generic implants, many women have opted to receive gender-specific knee implants (Fig. 6.5). Much like the data on generic TKRs, there is no significant effect on clinical outcome in women using gender-specific TKRs instead of generic TKRs; major evidence-based studies have shown similar radiologic outcome, range of motion, and functional scores [50, 65–68]. Interestingly, a recent study in which patients underwent a bilateral TKA receiving one gender-specific knee and one generic knee noted similar results. Patients preferred the generic and gender-specific knees at the same rates [67].

Final Remarks

The data is essentially unequivocal in finding no advantage for gender-specific knee implants and no difference in clinical outcome in women with standard generic knee implants. It is worth noting, however, that studies on gender-specific knee implants are early results due to their recent introduction. Knee implants are expected to last 20–30 years, so it is unknown whether these gender-specific knee implants will last longer or shorter in women than conventional total knee replacements.

Hip

Generic and Gender-Specific Total Hip Replacement (THR)

To date, there are no gender-specific hip replacements on the market. However, there are custom hips to fit the anatomy of individual patients in certain situations.

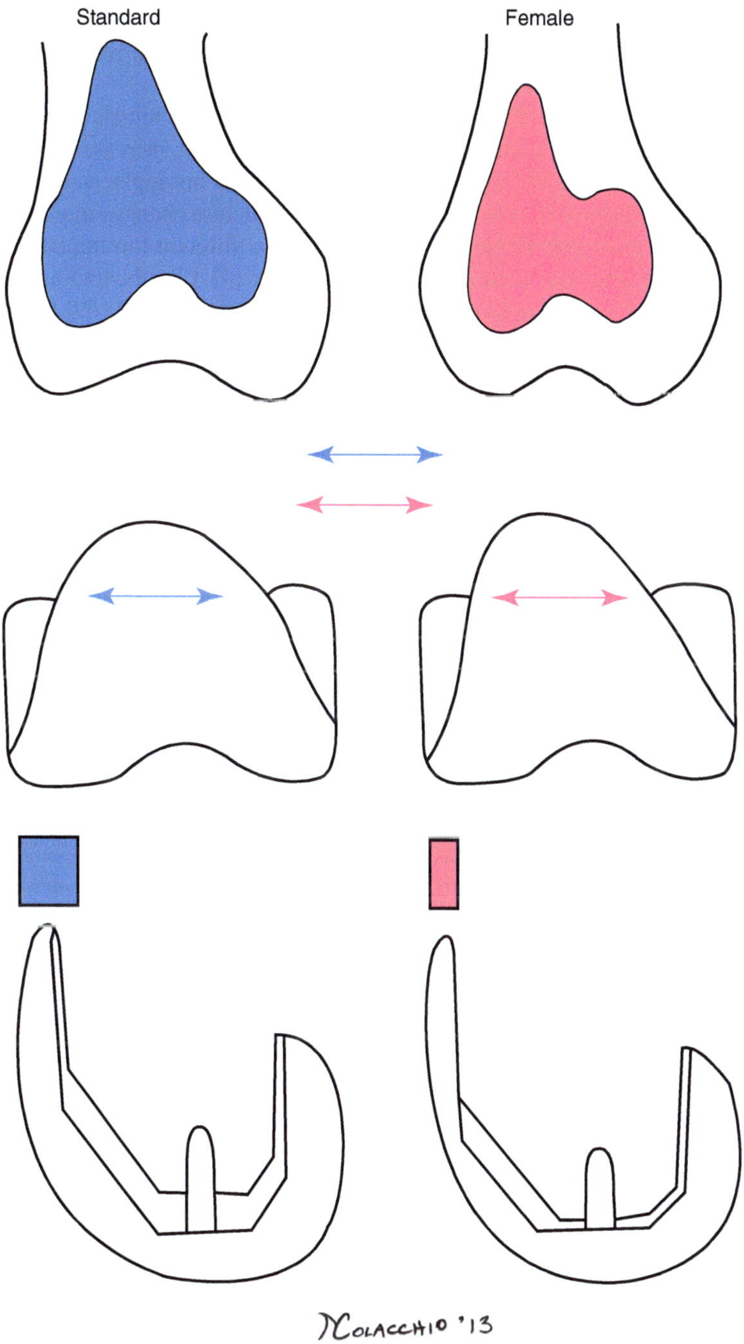

Fig. 6.5 Example of a gender-specific knee implant (Zimmer)

Research has primarily focused on determining whether the anatomical differences suggest a need for gender-specific THR and whether the current standard hip implants work as well in women as they do in men.

As mentioned earlier, clear anatomical differences in the female hip have been demonstrated; a smaller acetabulum, a shorter femoral head, increased anteversion, and a larger Q angle could impact the outcome of total hip replacement. There is also data that suggests as women age, their bone structure changes more than men [52], which implies that women may benefit from a different hip implant because this is a surgery generally performed on older patients [4]. Like the findings in TKR, women have higher pain and lower functional ability prior to THR [69]. The AAOS has recommended, based on female anatomy, female aging, biomechanics, and the female burden of osteoporosis, the production of a hip implant for women with a femoral stem that has a smaller metaphysis and shorter base neck [52].

Despite anatomic and biomechanical indications for gender-specific THR, the studies on outcomes between women and men with standard THR generally suggest no need for the use of gender-specific THR. A major review of the THR literature by the Clinical Orthopaedics and Related Research journal concluded that standard THR systems, which already have the capability to adjust for slight anatomic differences, have not led to different outcomes between men and women [50]. They do *not* see the benefit of developing and using gender-specific total hip implants if the standard hip implants are sufficient.

There have been some peer-reviewed studies in isolation that could suggest a need for gender-specific hip implants or at least some revision of the current hip implant protocol for women. Women have been shown to be at higher risk than men for peri-prosthetic fracture after THR [70], abnormal gait 1 year after THR [71], and for increased pain, NSAID use, and narcotic use 2–5 years after THR [72]. In addition, a low bone muscle density (as seen in osteoporosis) has been shown to lead to slower femoral stem osseointegration and poorer initial stability in women [73].

Final Remarks

There are currently no hip replacements designed specifically for women, and the evidence is still unclear as to whether a gender-specific hip is necessary. Anatomic differences suggest a potential use for them, but if current THRs are sufficient, then maybe it is not necessary for manufacturers to create a “fix” to a problem that does not exist. Women should generally feel very comfortable receiving a standard hip replacement.

Conclusion

Osteoarthritis is a very prevalent and debilitating disease that affects women more than men, although the exact reasons for this predilection remain unclear. There are many treatments for osteoarthritis that should be exhausted before opting for surgery. Total joint replacement is a safe and effective procedure to relieve knee and hip pain and improves functionality resulting from osteoarthritis. Results in women have been on par with the results observed in men using both gender-neutral and gender-specific joint implants, and thus, it does not seem necessary for women to seek more costly and less proven gender-specific implants.

References

1. Bijlsma JW, Berenbaum F, Lafeber FP. Osteoarthritis: an update with relevance for clinical practice. Lancet. 2011;377(9783):2115–26.
2. Lawrence RC, Felson DT, Helmick CG, et al. Estimates of the prevalence of arthritis and other rheumatic conditions in the united states. Part II. Arthritis Rheum. 2008;58(1):26–35.
3. Brault MW, Hootman J, Helmick CG, Theis KA, Armour BS. Prevalence and most common causes of disability among adults. MMWR Morb Mortal Wkly Rep. 2009;58(16):421.
4. Healthcare cost and utilization project (HCUP), Nationwide inpatient sample (NIS). Agency for healthcare research and quality: Rockville. http://hcupnet.ahrq.gov/2009. Accessed Dec 2012.
5. O'Connor MI, Hooten EG. Breakout session: gender disparities in knee osteoarthritis and TKA. Clin Orthop Relat Res. 2011;469(7):1883–5.
6. Srikanth VK, Fryer JL, Zhai G, Winzenberg TM, Hosmer D, Jones G. A meta-analysis of sex differences prevalence, incidence and severity of osteoarthritis. Osteoarthritis Cartilage. 2005; 13(9):769–81.
7. Cho H, Chang C, Yoo J, Kim S, Kim T. Gender differences in the correlation between symptom and radiographic severity in patients with knee osteoarthritis. Clin Orthop Relat Res. 2010;468(7):1749–58.
8. Keefe FJ, Lefebvre JC, Egert JR, Affleck G, Sullivan MJ, Caldwell DS. The relationship of gender to pain, pain behavior, and disability in osteoarthritis patients: the role of catastrophizing. Pain. 2000;87(3):325–34.
9. Tosi LL, Boyan BDP, Boskey ALP. Does sex matter in musculoskeletal health? The influence of sex and gender on musculoskeletal health. J Bone Joint Surg Am. 2005;87-A(7):1631–47.
10. Tak SH, Laffrey SC. Life satisfaction and its correlates in older women with osteoarthritis. Orthop Nurs. 2003;22(3):182–9.
11. Hanna FS, Teichtahl AJ, Wluka AE, et al. Women have increased rates of cartilage loss and progression of cartilage defects at the knee than men: a gender study of adults without clinical knee osteoarthritis. Menopause. 2009;16(4):666–70.
12. Tsai CL, Liu TK. Osteoarthritis in women: its relationship to estrogen and current trends. Life Sci. 1992;50(23):1737–44.
13. Stevens-Lapsley JE, Kohrt WM. Osteoarthritis in women: effects of estrogen, obesity and physical activity. Womens Health (Lond Engl). 2010;6(4):601–15.
14. Lian K, Lui L, Zmuda JM, et al. Estrogen receptor alpha genotype is associated with a reduced prevalence of radiographic hip osteoarthritis in elderly Caucasian women. Osteoarthritis Cartilage. 2007;15(8):972–8.
15. Fytili P, Giannatou E, Papanikolaou V, et al. Association of repeat polymorphisms in the estrogen receptors alpha, beta, and androgen receptor genes with knee osteoarthritis. Clin Genet. 2005;68(3):268–77.
16. Sharma L, Song J, Felson DT, Cahue S, Shamiyeh E, Dunlop DD. The role of knee alignment in disease progression and functional decline in knee osteoarthritis. JAMA. 2001;286(2):188–95.
17. Felson DT. Clinical practice. Osteoarthritis of the knee. N Engl J Med. 2006;354(8):841–8.
18. Moore K, Dalley A, Agur A. Clinically oriented anatomy. 6th ed. Baltimore: Lippincott Williams & Wilkins; 2010. p. 331.

19. Grelsamer RP, Dubey A, Weinstein CH. Men and women have similar Q angles: a clinical and trigonometric evaluation. J Bone Joint Surg Br. 2005;87(11):1498–501.
20. Woodland LH, Francis RS. Parameters and comparisons of the quadriceps angle of college-aged men and women in the supine and standing positions. Am J Sports Med. 1992;20(2):208–11.
21. Chmell M, McManus J, Scott R. Thickness of the patella in men and women with osteoarthritis. Knee. 1995;2(4):239–41.
22. Jones G, Glisson M, Hynes K, Cicuttini F. Sex and site differences in cartilage development: a possible explanation for variations in knee osteoarthritis in later life. Arthritis Rheum. 2000; 43(11):2543–9.
23. Prodromos CC, Han Y, Rogowski J, Joyce B, Shi K. A meta-analysis of the incidence of anterior cruciate ligament tears as a function of gender, sport, and a knee injury-reduction regimen. Arthroscopy. 2007;23(12):1320–1325.e6.
24. Arendt E, Dick R. Knee injury patterns among men and women in collegiate basketball and soccer. NCAA data and review of literature. Am J Sports Med. 1995;23(6):694–701.
25. Vairo GL, McBrier NM, Miller SJ, Buckley WE. Premature knee osteoarthritis after anterior cruciate ligament reconstruction dependent on autograft. J Sport Rehabil. 2010;19(1): 86–97.
26. Hootman JM, Dick R, Agel J. Epidemiology of collegiate injuries for 15 sports: summary and recommendations for injury prevention initiatives. J Athl Train. 2007;42(2):311–9.
27. Sridhar MS, Jarrett CD, Xerogeanes JW, Labib SA. Obesity and symptomatic osteoarthritis of the knee. J Bone Joint Surg Br. 2012;94(4):433–40.
28. Teichtahl AJ, Wluka AE, Proietto J, Cicuttini FM. Obesity and the female sex, risk factors for knee osteoarthritis that may be attributable to systemic or local leptin biosynthesis and its cellular effects. Med Hypotheses. 2005;65(2):312–5.
29. Cicuttini FM, Baker JR, Spector TD. The association of obesity with osteoarthritis of the hand and knee in women: a twin study. J Rheumatol. 1996;23(7):1221–6.
30. Grotle M, Hagen KB, Natvig B, Dahl FA, Kvien TK. Obesity and osteoarthritis in knee, hip and/or hand: an epidemiological study in the general population with 10 years follow-up. BMC Musculoskelet Disord. 2008;9:132-2474-9-132.
31. Reijman M, Pols HA, Bergink AP, et al. Body mass index associated with onset and progression of osteoarthritis of the knee but not of the hip: the Rotterdam study. Ann Rheum Dis. 2007; 66(2):158–62.
32. Jackson AS, et al. The effect of sex, age and race on estimating percentage body fat from body mass index: the heritage family study. Int J Obes Relat Metab Disord. 2002;26(6):789–96.
33. Karvonen-Gutierrez CA, Sowers MR, Heeringa SG. Sex dimorphism in the association of cardiometabolic characteristics and osteophytes-defined radiographic knee osteoarthritis among obese and non-obese adults: NHANES III. Osteoarthritis Cartilage. 2012;20(7): 614–21.
34. International osteoporosis foundation (IOF): Who's at risk? http://www.iofbonehealth.org/whos-risk. Accessed March 2013.
35. National Institutes of Health Osteoporosis and Related Bone Diseases National Resource Center Osteoporosis and arthritis: two common but different conditions. 2012.
36. Labuda A, Papaioannou A, Pritchard J, Kennedy C, DeBeer J, Adachi JD. Prevalence of osteoporosis in osteoarthritic patients undergoing total hip or total knee arthroplasty. Arch Phys Med Rehabil. 2008;89(12):2373–4.
37. Makinen TJ, Alm JJ, Laine H, Svedstrom E, Aro HT. The incidence of osteopenia and osteoporosis in women with hip osteoarthritis scheduled for cementless total joint replacement. Bone. 2007;40(4):1041–7.
38. Clinical Practice Guideline Work Group. American Academy of Orthopaedic Surgeons (AAOS). Rosemont, IL 60018: American Academy of Orthopaedic Surgeons Clinical Practice Guideline on the Treatment of Osteoarthritis of the Knee 2013.
39. Hochberg MC, Altman RD, April KT, et al. American college of rheumatology 2012 recommendations for the use of nonpharmacologic and pharmacologic therapies in osteoarthritis of the hand, hip, and knee. Arthritis Care Res (Hoboken). 2012;64(4):455–74.

40. Dennison EM, Arden NK, Kellingray S, Croft P, Coggon D, Cooper C. Hormone replacement therapy, other reproductive variables and symptomatic hip osteoarthritis in elderly white women: a case–control study. Br J Rheumatol. 1998;37(11):1198–202.
41. Marjoribanks J, Farquhar C, Roberts H, Lethaby A. Long term hormone therapy for perimenopausal and postmenopausal women. Cochrane Database Syst Rev. 2012;7, CD004143.
42. Von Muhlen D, Morton D, Von Muhlen CA, Barrett-Connor E. Postmenopausal estrogen and increased risk of clinical osteoarthritis at the hip, hand, and knee in older women. J Womens Health Gend Based Med. 2002;11(6):511–8.
43. Altman RD, Fowler PJ. Pharmacologic treatment of knee osteoarthritis in athletic women. Phys Sportsmed. 2011;39(3):39–44.
44. Silva A, Serrao PR, Driusso P, Mattiello SM. The effects of therapeutic exercise on the balance of women with knee osteoarthritis: a systematic review. Rev Bras Fisioter. 2012;16(1):1–9.
45. Kurtz S, Ong K, Lau E, Mowat F, Halpern M. Projections of primary and revision hip and knee arthroplasty in the united states from 2005 to 2030. J Bone Joint Surg Am. 2007;89(4): 780–5.
46. Lingard EA, Katz JN, Wright EA, Sledge CB, Kinemax Outcomes Group. Predicting the outcome of total knee arthroplasty. J Bone Joint Surg Am. 2004;86-A(10):2179–86.
47. Januel JM, Chen G, Ruffieux C, et al. Symptomatic in-hospital deep vein thrombosis and pulmonary embolism following hip and knee arthroplasty among patients receiving recommended prophylaxis: a systematic review. JAMA. 2012;307(3):294–303.
48. Vissers MM, Bussmann JB, Verhaar JA, Arends LR, Furlan AD, Reijman M. Recovery of physical functioning after total hip arthroplasty: systematic review and meta-analysis of the literature. Phys Ther. 2011;91(5):615–29.
49. Lu ZH, Yu JK, Chen LX, Gong X, Wang YJ, Leung KK. Computed tomographic measurement of gender differences in bowing of the sagittal femoral shaft in persons older than 50 years. J Arthroplasty. 2012;27(6):1216–20.
50. Johnson AJ, Costa CR, Mont MA. Do we need gender-specific total joint arthroplasty? Clin Orthop Relat Res. 2011;469(7):1852–8.
51. Poilvache PL, Insall JN, Scuderi GR, Font-Rodriguez DE. Rotational landmarks and sizing of the distal femur in total knee arthroplasty. Clin Orthop Relat Res. 1996;331(331):35–46.
52. Glassman A, Lachiewicz P, Tanzer M. Hip and knee reconstruction. In: Orthopaedic knowledge update. 4th ed. Rosemont: American Academy of Orthopaedic Surgeons; 2011.
53. Mahomed NN, Barrett J, Katz JN, Baron JA, Wright J, Losina E. Epidemiology of total knee replacement in the United States Medicare population. J Bone Joint Surg Am. 2005;87(6): 1222–8.
54. Hawker GA, Wright JG, Coyte PC, et al. Differences between men and women in the rate of use of hip and knee arthroplasty. N Engl J Med. 2000;342(14):1016–22.
55. Petterson SC, Raisis L, Bodenstab A, Snyder-Mackler. Disease-specific gender differences among total knee arthroplasty candidates. J Bone Joint Surg Am. 2007;89–A(11):2327–33.
56. Tonelli SM, Rakel BA, Cooper NA, Angstom WL, Sluka KA. Women with knee osteoarthritis have more pain and poorer function than men, but similar physical activity prior to total knee replacement. Biol Sex Differ. 2011;2:12.
57. O'Connor MI. Implant survival, knee function, and pain relief after TKA: Are there differences between men and women? Clin Orthop Relat Res. 2011;469(7):1846–51.
58. Lin J, Yang B, Weng XS, Jin J, Zhao Q, Qiu GX. Effect of osteoarthritis patients' gender on rehabilitation after total knee arthroplasty. Chin Med Sci J. 2009;24(2):102–6.
59. Pun SY, Ries MD. Effect of gender and preoperative diagnosis on results of revision total knee arthroplasty. Clin Orthop Relat Res. 2008;466(11):2701–5.
60. Ritter MA, Wing JT, Berend ME, Davis KE, Meding JB. The clinical effect of gender on outcome of total knee arthroplasty. J Arthroplasty. 2008;23(3):331–6.
61. Lustig S, Barba N, Magnussen RA, Servien E, Demey G, Neyret P. The effect of gender on outcome of unicompartmental knee arthroplasty. Knee. 2012;19(3):176–9.
62. Sensi L, Buzzi R, Giron F, De Luca L, Aglietti P. Patellofemoral function after total knee arthroplasty: gender-related differences. J Arthroplasty. 2011;26(8):1475–80.

63. Kamath AF, Horneff JG, Gaffney V, Israelite CL, Nelson CL. Ethnic and gender differences in the functional disparities after primary total knee arthroplasty. Clin Orthop Relat Res. 2010;468(12):3355–61.
64. Liebs T, Herzberg W, Roth-Kroeger A, Rüther W, Hassenpflug J. Women recover faster than men after standard knee arthroplasty. Clin Orthop Relat Res. 2011;469(10):2855–65.
65. Kim Y, Choi Y, Kim J. Comparison of a standard and a gender-specific posterior cruciate-substituting high-flexion knee prosthesis: a prospective, randomized, short-term outcome study. J Bone Joint Surg Am. 2010;92(10):1911–20.
66. Tanavalee A, Rojpornpradit T, Khumrak S, Ngarmukos S. The early results of gender-specific total knee arthroplasty in Thai patients. Knee. 2011;18(6):483–7.
67. Song EK, Jung WB, Yoon TR, Park KS, Seo HY, Seon JK. Comparison of outcomes after bilateral simultaneous total knee arthroplasty using gender-specific and unisex knees. J Arthroplasty. 2012;27(2):226–31.
68. Kim YH, Choi Y, Kim JS. Comparison of standard and gender-specific posterior-cruciate-retaining high-flexion total knee replacements: a prospective, randomised study. J Bone Joint Surg Br. 2010;92(5):639–45.
69. Holtzman J, Saleh K, Kane R. Gender differences in functional status and pain in a Medicare population undergoing elective total hip arthroplasty. Med Care. 2002;40(6):461–70.
70. Singh JA, Jensen MR, Harmsen SW, Lewallen DG. Are gender, comorbidity, and obesity risk factors for postoperative periprosthetic fractures after primary total hip arthroplasty? J Arthroplasty. 2013;28(1):126–31.
71. Perron M, Malouin F, Moffet H, McFadyen BJ. Three-dimensional gait analysis in women with a total hip arthroplasty. Clin Biomech (Bristol, Avon). 2000;15(7):504–15.
72. Singh JA, Lewallen D. Age, gender, obesity, and depression are associated with patient-related pain and function outcome after revision total hip arthroplasty. Clin Rheumatol. 2009;28(12): 1419–30.
73. Aro HT, Alm JJ, Moritz N, Makinen TJ, Lankinen P. Low BMD affects initial stability and delays stem osseointegration in cementless total hip arthroplasty in women: a 2-year RSA study of 39 patients. Acta Orthop. 2012;83(2):107–14.

Chapter 7
Nutraceuticals: An Alternative for Osteoarthritis Management

Emily J. Curry, Jennifer Baima, and Elizabeth Matzkin

Abstract Dietary supplements, known as nutraceuticals, are commonly used to treat medical conditions in the United States. Despite widespread use, it is unnecessary for companies to seek FDA approval before marketing and manufacturing these supplements. Nutraceuticals are typically thought to have medicinal properties, but also may consist of nutrients extracted from commonly consumed foods. Osteoarthritis prevention and treatment is the most common reason for nutraceutical supplementation and use is expected to increase with the aging "baby boomer" pop ulation. Physician and pharmacist consultation is essential for patient safety, given the side effects and toxicity associated with some nutraceuticals. A basic understanding of currently available supplements and their side effects is crucial in patient care today. This chapter will discuss the currently available nutraceuticals for the treatment of osteoarthritis and the supporting literature for each supplement. Supplements discussed will include glucosamine, chondroitin, omega-3 fatty acids, avocado-soybean unsaponifiables, vitamins (vitamin A, vitamin C, and vitamin E), minerals (boron, zinc, copper), and willow bark. None of these common nutraceuticals have substantial evidence of efficacy in prevention or progression of osteoarthritis.

Keywords Nutraceuticals • Osteoarthritis • Vitamins • Minerals • Dietary supplements • Omega-3 fatty acids • Glucosamine • Chondroitin • Avocado-soybean unsaponifiables

E.J. Curry, BA • E. Matzkin, MS, MD (✉)
Department of Orthopedic Surgery, Brigham and Women's Hospital,
75 Francis Street, Boston, MA 02115, USA
e-mail: ecurry1@partners.org; ematzkin@partners.org

J. Baima, MD
Division of Women's Health, Department of Medicine,,
Harvard Medical School, Brigham and Women's Hospital,
Chestnut Hill, MA 02467, USA
e-mail: jbaima1@partners.org

E. Mody, E. Matzkin (eds.), *Musculoskeletal Health in Women*,
DOI 10.1007/978-1-4471-4712-1_7,

Abbreviations

AAOS	American Academy of Orthopaedic Surgeons
ADR	Adverse drug reaction
ALA	Alpha-linolenic acid
ASU	Avocado-soybean unsaponifiable
BMI	Body mass index
COMP	Cartilage oligomeric matrix protein
COX	Cyclooxygenase pathway
DHA	Docosahexaenoic acid
EPA	Eicosapentaenoic acid
FA	Fatty acids
FDA	Federal Drug Administration
GAIT	Glucosamine and Chondroitin Arthritis Intervention Trial
IU	International unit
LOX	Lipoxygenase pathway
MRI	Magnetic Resonance Imaging
MSM	Methylsulfonylmethane
NIH	National Institute of Health
NSAID	Nonsteroidal anti-inflammatory drug
OA	Osteoarthritis
PUFA	Polyunsaturated fatty acids
TGF-beta	Transforming growth factor-beta

Introduction

Dietary supplements are commonly used to treat medical conditions in the United States. Over 100 million Americans consume dietary supplements, yet it is unnecessary for companies to seek FDA approval before marketing and manufacturing these supplements due to the Dietary Supplement Health and Education Act that was passed in 1994 [1, 2]. Based on this ruling, FDA marketing intervention will only occur if a supplement is deemed unsafe for use, but does not require evidence for efficacy if the supplement was sold in the United States before 1994 [1]. In contrast, developing a pharmaceutical drug with FDA approval takes a company nearly 10 years and can cost between 0.8 to 1.7 billion dollars for research, development, and marketing [3]. Nutraceutical development and marketing requires minimal effort by comparison. As a result of these minimal regulations, the nutraceutical field is now a 28 billion dollar industry annually [2].

"Osteoarthritis" prevention and treatment is the number one reason for dietary supplementation with over 50 % of nutraceuticals used for musculoskeletal pain [4]. The most common consumer demographics are educated females over 50 years of age with a normal BMI [4]. Physician and pharmacist consultation is essential for patient safety, given the side effects and toxicity associated with some nutraceuticals.

Over 50 % of patients in the United Kingdom consult a general practitioner about nutraceuticals before starting a formal regimen [5]. A basic understanding of currently available supplements and their side effects is crucial in patient care today.

DeFelice first coined the term nutraceutical in 1989 by combining the terms "nutrition" and "pharmaceutical" [6]. Nutraceuticals are generally considered to have medicinal properties; however, controversy remains as to whether commonly consumed foods in regular daily diets are also considered nutraceuticals. For example, avocado and soybeans are commonly consumed foods, yet extracts are also thought to treat osteoarthritis. Although nutraceuticals lack a formal definition with consensus, Zeisel described them as "diet supplements that deliver a concentrated form of a presumed bioactive agent from a food presented in a nonfood matrix, and used to enhance health in dosages that exceed those that could be obtained from normal foods" [7]. This chapter will discuss the currently available nutraceuticals for the treatment of osteoarthritis and the supporting literature for each supplement.

Glucosamine and Chondroitin

Glucosamine and chondroitin are two of the most commonly available supplements used for the treatment of osteoarthritis in the United States with sales exceeding 810 million dollars in 2005 [4, 8]. Glucosamine and chondroitin are cartilage and synovial fluid extracellular matrix constituents. They are naturally synthesized by chondrocytes and synoviocytes but also can be orally supplemented [9]. Glucosamine typically is derived from shellfish chitin, while chondroitin sulfate is commonly derived from bovine or shark cartilage [10, 11]. Therefore, patients with a shellfish allergy or iodine hypersensitivity should not consider glucosamine use. Glucosamine is available with sulfate or hydrochloride for optimal pharmacokinetic administration. While the mechanism of action for glucosamine and chondroitin is unknown, a series of hypotheses exist. Although mediated through different pathways, both may enhance chondrocyte synthesis by stimulating proteoglycan production, reduce proteolytic effects on chondrocytes, and reduce expression of inflammatory factors [12–15].

Controversy over glucosamine and chondroitin efficacy exists in the literature. The most widely cited multicenter trial is the GAIT trial (Glucosamine and Chondroitin Arthritis Intervention Trial) funded by the NIH in 2006 [16]. Over 1,500 patients with knee osteoarthritis were randomly given glucosamine, chondroitin sulfate, glucosamine and chondroitin sulfate, celecoxib, or placebo daily for 24 weeks. Overall, patients did not have significant pain relief or reduced disease progression from any glucosamine or chondroitin sulfate combination as compared to the placebo. The GAIT trial has limitations. Most patients selected for this study only had mild osteoarthritis and, therefore, did not have room for significant improvement [17].

Glucosamine supplementation has also been assessed during physical activity. Petersen et al. found that glucosamine significantly decreased serum cartilage oligomeric matrix protein (COMP), an indicator of cartilage turnover, over a 12-week strength training regimen [18]. Supplementation with glucosamine was also found to improve muscle strength during resistance training for patients with osteoarthritis;

however, muscle mass gain appears to remain unaffected [19, 20]. In contrast, obese women with knee OA had minimal functional improvements when treated with a combination of exercise and glucosamine/chondroitin and MSM supplementation in addition to a supervised weight loss program [21].

Several studies have evaluated the efficacy of pain relief with glucosamine supplements compared to Tylenol and oral anti-inflammatories [22–24]. Patients often consider alternative treatments for pain relief, since long-term use of acetaminophen (Tylenol) may lead to increased risk for liver and kidney failure [25, 26]. These risks are increased among the older patient populations and since 20 % (70 million) of Americans will be 65 or older by 2030; this problem may only magnify over time [27]. There is some evidence to support that glucosamine has a delayed pain relief effect compared to ibuprofen. Two studies showed that ibuprofen provides superior short-term pain relief within the first 2 weeks of supplementation, but glucosamine becomes equal or more effective in the long term [22, 23, 28]. Similarly, chondroitin was found to have significantly better long-term pain relief effects than diclofenac [24]. These studies suggest that oral glucosamine and chondroitin may be safer and possibly more effective than NSAIDS or acetaminophen for osteoarthritis pain relief.

While controversy still exists over glucosamine/chondroitin sulfate efficacy, both supplements are considered safe for use with very few adverse events. Glucosamine is safe for use among patients with Type II diabetes despite containing a monosaccharide constituent. Safety profile was confirmed in both Type II diabetics and normal subjects, where the use of glucosamine /chondroitin did not elevate insulin and blood glucose levels after 3 months of use [29, 30]. Although rare, the most common side effects of glucosamine include epigastric pain (3.5 %), heartburn (2.7 %), and diarrhea (2.5 %) [31]. Incidence and type of side effects are similar between glucosamine or chondroitin and placebo [32, 33].

The major limitations of glucosamine and chondroitin studies are secondary to the chemical inclusions (hydrochloride or sulfate), which have a direct effect on bioavailability and absorption [34]. Reduced bioavailability within the GAIT study may explain the unfavorable results, since glucosamine hydrochloride (less bioavailability when compared to glucosamine sulfate) was used in all patients. In 2008, Clinical Practice Guidelines published by the AAOS recommended against the use of glucosamine and chondroitin for knee osteoarthritis treatment due to a lack of high-level evidence showing clinical efficacy [35]. Current recommendations suggest a limited role of glucosamine and chondroitin usage as a long-term pain relief in patients with osteoarthritis; however, the safety profile has been well studied and literature supports that long-term use of glucosamine and chondroitin will not likely result in any major side effects.

Omega-3 PUFAs

Since the human body cannot synthesize omega-3 and omega-6 polyunsaturated fatty acids (PUFAs), dietary supplementation is crucial. Omega-3 PUFAs commonly found in flaxseed, canola oil, and fish oil have known anti-inflammatory benefits for cardiovascular and inflammatory diseases. In contrast, omega-6 fatty acids (FAs), commonly found in corn oil and meat, have characteristics that can counteract the positive effects of omega-3 supplementation. Since the American diet now has an increased quantity of meat and corn by-products, the dietary ratio of omega-6 to omega-3 FAs is now close to 20:1, instead of the recommended 4:1 or 1:1 [36, 37]. Adjusting FA consumption has beneficial effects far beyond the potential prevention of osteoarthritis progression, such as coronary plaque and cardiovascular disease prevention. Omega-3 use may benefit patients with rheumatoid arthritis; however, controversy remains over whether omega-3 is effective in the treatment of osteoarthritis [38, 39].

In vitro studies assessing omega-3 supplementation for the treatment of osteoarthritis show promising results secondary to omega-3 precursor's participation in the COX and LOX anti-inflammatory enzyme pathway, which ultimately produces eicosanoids and docosanoids [40]. However, current in vivo data remains mixed. In the MOST study, increased omega-3 presence correlated with increased patellofemoral cartilage presence, but not for tibiofemoral cartilage [41]. Furthermore, a higher ratio of omega-6 compared to omega-3 was correlated with the presence of synovitis on MRI. This study suggests that omega-3 fatty acid supplementation may have protective effects against cartilage loss and synovitis progression [41–44]. In contrast, two older trials done by Stammers et al. comparing omega-3 (eicosapentaenoic acid or EPA) to placebo did not show significantly decreased pain or increased function in patients with OA [42, 43].

Omega-3 supplements are derived from plant-based (ALA) or marine-based (EPA or DHA) sources. The body cannot convert ALA into EPA or DHA as easily for anti-inflammatory precursors; so the marine-based derivatives are the optimal form for supplementation [40]. Krill oil, in particular, is thought to improve OA symptoms as a result of the antioxidant, astaxanthin, which is naturally found within the oil [44]. Due to promising overall health benefits and few contraindications, omega-3 PUFA supplementations remain an interesting alternative; however, evidence for OA symptom reduction and prevention of progression remains controversial.

Avocado-Soybean Unsaponifiables (ASUs)

The anti-inflammatory and anabolic properties of avocado-soybean unsaponifiables (ASUs) may be useful in patients with osteoarthritis; however, higher levels of evidence are lacking to support its efficacy in the treatment of OA. ASUs have been shown to inhibit inflammatory cytokines, including PGE2, IL-1, IL-6, and IL-8 [45, 46]. ASUs have also been shown to stimulate collagen and aggrecan synthesis and decrease spontaneous collagenase activity [45, 46]. The most common ASU called Piascledine 300® (Parmascience, Inc, Montreal, Quebec, Canada) is available by prescription and is composed of two-thirds soybean oil and one-third avocado. Although ASUs are also available over the counter, Piascledine 300® is the standard pharmaceutical grade ASU used in all OA clinical trials due to its high-quality refining methods [47]. Piascledine 300® is most commonly prescribed in France.

The in vitro and in vivo results vary for ASU's role in OA symptom reduction. Three randomized double blind-studies evaluating the efficacy of ASUs reported with a 3–6-month supplementation period that patients with OA had decreased NSAID use for pain control in the ASU-supplemented group compared to placebo [48–50]. Two of these studies also found significantly increased function [49, 50]. However, statistically significant decrease in pain was only found in one study [49]. Maheu et al. found reduced joint space narrowing compared to the placebo with ASU supplementation over a 3-year period [51]. Lequesne et al. also found reduced joint space narrowing over a 2-year period; however, this study did not show significantly reduced pain, NSAID use, or increased function compared to placebo [52]. Furthermore, increasing dosage from 300 to 600 mg did not have an increased benefit to patients [48]. In contrast to the above studies, a placebo-controlled study by Walker-Bone et al. [53] failed to show any radiographic or functional improvement when ASU supplements were taken for a 2-year time period. However, this study did show that patients with severe hip OA taking Piascledine 300® had significantly reduced joint space narrowing compared to placebo. A recent meta-analysis concluded ASUs are effective in improving function in patients with OA and was found to have "good evidence of efficacy" [54]. This meta-analysis also concluded that patients with knee OA benefit more than hip OA [53, 54]. Based on the current literature, it appears that ASUs may play a role in decreasing rate of joint space narrowing in patients with OA, but ASU's role in long-term pain relief remains inconclusive.

Overall, patients with osteoarthritis tolerate ASUs well without significant side effects. According to the "French Spontaneous Reporting System," adverse drug reactions from Piascledine 300® were "very rare." Out of 117 ADRs reported in a 30-year period, women suffered adverse reactions more often than men and the most common side effects included eczema, urticarial, hepatic injury, colitis, and diarrhea [55]. Based on the currently available literature, patients with osteoarthritis may use ASU for treatment since the side effects are overall mild; however, patients should be aware that there is currently no substantial evidence to support its use. Furthermore, most studies are done with the standardized Piascledine 300®, yet a wide range of ASUs are available on the market from different companies with varying concentrations of avocado and soybean extracts.

Vitamins

Vitamins have known antioxidant properties that aid in reducing free radicals and may potentially reduce osteoarthritis progression. Most patients explore nutraceutical supplementation alternatives to NSAID use secondary to the potential side effects associated with NSAIDs. Although vitamins do have antioxidant properties to support various preventative measures, excessive supplementation can have deleterious effects. Patients interested in self-prescribing nutraceuticals should use caution when boosting vitamin A, C, or E intake without consulting a physician due to the potential side effects and possible drug interactions.

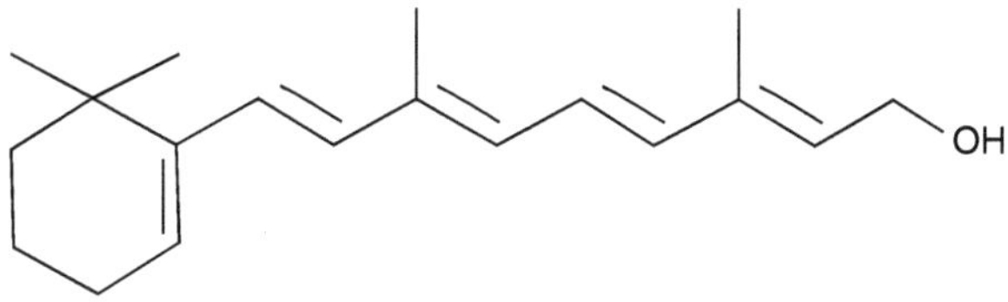

Fig. 7.1 Chemical structure of vitamin A (Used with permission from Nelson and Cox [95])

Vitamin A

Vitamin A (Fig. 7.1), also known as retinol, is a derivative of beta-carotene and is readily consumed in the human diet. Beta-carotene is one of four naturally occurring carotenoids. Beta-carotene is converted into vitamin A within the body and is commonly found in foods, such as carrots, kale, spinach, and sweet potatoes. No randomized placebo-controlled trials currently exist in the literature to assess the role of vitamin A deficiency in osteoarthritis progression. Furthermore, a systematic review by Canter et al. concluded that vitamin A use to prevent OA progression is inconclusive and doubtful [56]. Only one randomized control trial assessed vitamin A efficacy in OA-related pain relief and found that vitamin A supplementation was an insignificant mediator of pain relief. However, this study was underpowered and, furthermore, lacked a placebo control group, and the patients in this study already had a normal serum level of vitamin A. The study concluded that vitamin A supplementation alone was not successful in pain relief [57]. Conflicting results should also be considered when assessing serum carotenoid levels and knee osteoarthritis progression. De Roos et al. concluded that subjects with higher serum levels of lutein or beta-cryptoxanthin were less likely to have degenerative radiographic changes, whereas patients with high levels trans-beta-carotene and zeaxanthin were more likely to have radiograph-confirmed osteoarthritic changes [58]. Therefore, serum carotenoid type may be a contributor to osteoarthritic progression; however, no Level I or Level II studies exist to assess this possible correlation. The revolutionary Framingham Cohort was assessed for dietary beta-carotene levels and no association was found between decreased beta-carotene consumption and osteoarthritis when vitamin C levels remained high [59].

Current literature remains inconclusive regarding vitamin A's relationship to osteoarthritis symptoms and progression. The largest concern for vitamin A use is the possible side effects associated with vitamin A hypervitaminosis (toxicity) resulting from improper patient self-directed supplementation. Vitamin A toxicity is seen more commonly with acute supplementation of a single dose in excess of one million IU, but can also occur chronically by adults consuming greater than 50,000 IU/day for a longer duration [60]. Vitamin A toxicity can cause increased bone resorption, increased intracranial pressure, and hypercalcemia that can eventually cause liver damage [60, 61]. Patients should be aware that vitamin A supplements often are packaged with a carotenoid blend of lutein and zeaxanthin to prevent macular degeneration. Lutein is associated with reduced osteoarthritic radiographic changes, and zeaxanthin is associated with increased osteoarthritic changes [58]. Since osteoarthritic progression appears to increase with zeaxanthin supplementation, use of this supplement should be avoided

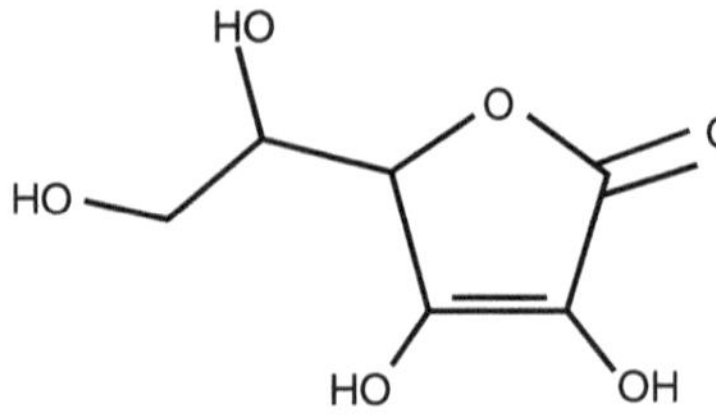

Fig. 7.2 Chemical structure of vitamin C (Nelson and Cox [95])

altogether. There is not enough evidence in the literature to support the use of vitamin A supplementation for OA pain relief and treatment; however, improper supplementation may contribute to vitamin A toxicity, which can have significant side effects for the patient. If a patient insists on using vitamin A supplementation, close monitoring, proper dosage, and physician-directed instructions are crucial.

Vitamin C

Vitamin C (Fig. 7.2) efficacy in treating osteoarthritis is inconclusive. Also known as ascorbic acid, vitamin C supplements are often found in combination with calcium, since both are crucial for overall bone health. Vitamin C aids in the synthesis of type II collagen and glycosaminoglycans [59, 62, 63]. Glycosaminoglycan synthesis or supplementation may be an important target in reducing osteoarthritic changes, since sulfated proteoglycan reduction is one of the first documented extracellular matrix alterations in early osteoarthritis [63]. However, vitamin C also plays a role in activating TGF-beta, which can worsen osteoarthritis symptoms [64]. As a result of both possible positive and negative effects, vitamin C use for osteoarthritis treatment is controversial.

A number of clinical trials suggest that vitamin C supplementation reduces pain in hip and knee osteoarthritis; however, an animal model supports a mechanism of action suggestive of TGF-beta activation, which would contribute negatively to OA progression. The Framingham Cohort Study from the 1990s concluded that 120–200 mg of vitamin C decreased pain and osteoarthritis progression risk [59]. Another study also showed decreased pain with 898 mg of vitamin C after supplementation for only 2 weeks [65]. A longitudinal study published in 2011 concludes that vitamin C does not have a protective role in knee OA progression but may play a role in overall incidence [66]. Unlike many other vitamins, vitamin C is water-soluble and has a low toxicity risk. The only documented side effect of excessive vitamin C consumption is diarrhea, which is rare.

While more research is needed to confirm or refute vitamin C's role in osteoarthritis, it is reasonable to assume that two mechanisms of action may be functioning in opposition to one another. While vitamin C supplementation may have other benefits, such as boosting the immune system, it should not be taken exclusively as a means to prevent osteoarthritis progression.

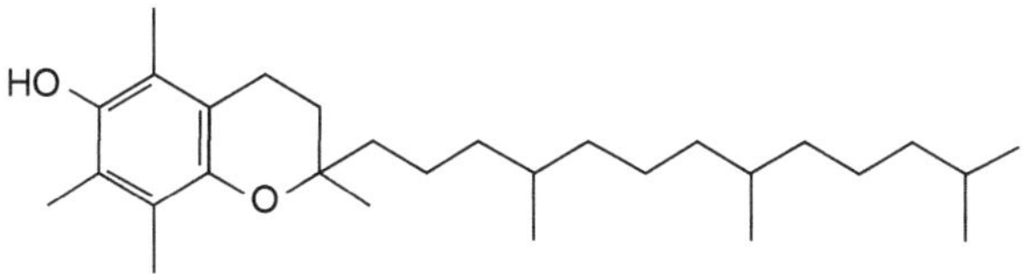

Fig. 7.3 Chemical structure of vitamin E (Nelson and Cox [95])

Vitamin E

Vitamin E (Fig. 7.3) is fat soluble and composed of both tocopherols and tocotrienols. Studies suggest that vitamin E may play a small role in osteoarthritis prevention due to its antioxidant properties. However, most studies in the literature are contradictory, and there is no high level evidence in the current literature suggesting vitamin E may prevent osteoarthritis prevention or symptom reduction. The Framingham Cohort Study concluded that men with greater vitamin E in the diet had less osteoarthritis progression, but did not play a role in overall incidence [59]. Furthermore, Bhattacharya et al. demonstrated vitamin E's role in reducing serum markers of oxidative stress-related OA progression and confirmed its efficacy; however, radiographic evidence and pain levels were not assessed [67].

In contrast, a 2-year randomized-controlled trial demonstrated no significant difference between vitamin E supplementation group and the placebo [68]. Most other studies assessing vitamin E lack placebo controls, are methodologically weak, or do not assess long-term vitamin E efficacy [56]. Although vitamin E efficacy in decreasing OA incidence and related pain is questionable at best, there are very few side effects in short-term supplementation. However, long-term supplementation may increase the risk of mortality due to heart failure [69]. Overall, there is very little supporting evidence that vitamin E reduces pain or OA progression either radiographically or symptomatically. Therefore, we cannot recommend the use of vitamin E for OA incidence, progression, or associated pain management.

Minerals

Given the paucity of data in the literature regarding mineral supplementation for osteoarthritis, few recommendations can be made in this arena. Boron is one of the most promising minerals; however, higher quality studies are needed to determine its overall efficacy. Of greater concern, many minerals have dangerous side effects as a result of toxicity. Unlike the vitamins discussed prior, mineral over-supplementation can sometimes result in other mineral deficiencies. Due to the side effects and precarious nature of dosing, mineral supplementation is not recommended. Boron, zinc, copper, and selenium are most commonly discussed in the osteoarthritis literature and a summary of the clinical findings for each is described next.

Boron

Boron is a promising mineral for osteoarthritis treatment, although many more clinical trials are needed to confirm its efficacy. Boron is typically found in high amounts in foods such as avocado, peanut butter, almonds, pecans, and prune juice [70]. Therefore, a diet deficient of fruits, vegetables, and nuts may be a cause for concern for a boron deficiency. Boron's mechanism of action is unknown, but current literature hypothesizes that boron supplementation may indirectly alter other mineral levels in the body, such as calcium and magnesium; alter hormone levels, such as estrogen; and alter reactive oxygen species presence [71, 72].

All proposed mechanisms for boron supplementation may contribute to osteoarthritis prevention. Similarly, the mechanism behind osteoarthritis prevention is unknown, but there is substantial evidence suggesting that boron deficiency may play a role in bone metabolism. Newnham et al. found that boron was two times less concentrated in osteoarthritic femoral heads as compared to healthy femoral heads [73]. However, this study was not randomized or placebo-controlled. This observational evidence prompted a pilot placebo-controlled study of 20 patients evaluating the efficacy of boron in the treatment of OA. The authors found that 50 % of patients with radiographically confirmed osteoarthritis improved with boron supplementation, whereas only 10 % improved with the placebo [74]. However, this study was limited by the sample size and thus underpowered.

There are currently no high-level studies assessing boron efficacy in osteoarthritis treatment. A randomized trial in a rabbit model concluded that boron supplementation enhanced bone strength and mineral concentration in conjunction with a high-energy diet [75]. Another small placebo-controlled single-blind study compared boron intake differences between sedentary women and female athletes [76]. Significantly higher bone mineral density was found in the active female athletes compared to sedentary females; however, this was not directly attributable to boron supplementation. However, this study did find that boron supplementation significantly increased blood magnesium and decreased phosphorus concentrations. Both magnesium and phosphorous contribute to bone metabolism.

Pure boron supplementation should be used with caution due to potential toxicity side effects. Consumption of boric acid in excess of 200 mg/kg is reported to be lethal [77]. Large doses may also result in symptoms such as riboflavonuria, tremors, convulsions, weakness, dermatitis, and hematemesis [78]. While boron certainly may play a role in bone metabolism, the precise benefit in patients with osteoporosis or who are at risk for developing osteoarthritis is inconclusive and supplementation cannot be recommended for this ailment alone. Instead, boron should be supplemented through fruit, vegetable, and nut sources to aid in the prevention of other potentially negative side effects of boron deficiency, such as poor mineral metabolism, impaired brain function, and decreased immune response [79, 80].

Zinc and Copper

Zinc and copper are often discussed together in osteoarthritis studies due to their relative concentration differences observed in osteoarthritic patients compared to normal patients. When zinc is consumed in high doses, copper concentration tends to decrease. Zinc is commonly found in cold and cough lozenges due to the purported benefit in reduced duration of upper respiratory symptoms. Zinc supplementation has been evaluated for osteoarthritis treatment. Serum zinc concentrations may be reduced in osteoarthritic patients, while copper concentrations may be increased [81]. Copper has also been found in significantly higher levels within synovial fluid of osteoarthritic joints [82], although a significant difference in zinc levels between plasma and synovial fluid concentrations has not been found.

Despite having this knowledge, the role of zinc and copper in osteoarthritis remains unknown and neither should be considered options in the treatment of osteoarthritis due to toxicity risk. For patients considering zinc supplementation without deficiency, "over-the-counter" supplementation can easily lead to overconsumption given that the therapeutic dose is defined as 5.2–16.2 mg, whereas most zinc supplements come in 25–50 mg doses recommended for a zinc deficiency [83]. Zinc overdose with10 g of supplementation or less has serious side effects, such as nausea, vomiting, and diarrhea. Furthermore, overconsumption leads to a copper deficiency that can contribute to neutropenia/anemia and impaired coordination [83]. Severe zinc toxicity can also cause pancreatitis, acute renal failure, and hemolysis. Zinc toxicity can even be fatal in a 10–30 g dose. Zinc and copper supplementation should not be considered for patients with osteoarthritis unless a preexisting deficiency exists.

Willow Bark

Willow bark, part of the Salix genus, often receives attention in treating musculoskeletal pain since it contains a component of aspirin called salicin [84]. Willow bark used as a nutraceutical is typically obtained from young trees in dried bark fragments [84]. The ancient Egyptians and Sumerians first discovered willow bark's analgesic and fever-reducing ability [85]. However, it was not until 1853 when Charles Gerhardt discovered acetylsalicylic acid could be extracted from the bark and was later formulated as "aspirin" by the Bayer company in 1897 [85]. Unlike aspirin, salicin does not irritate the stomach or impact blood coagulation as readily [86, 87]. However, willow bark extract consumption should not be advised in patients with an aspirin allergy due to reports of anaphylaxis and rash when consumed [88, 89]. Furthermore, willow bark extract should not be given to children to prevent the risk of Reye's syndrome secondary to the salicin [90].

Willow bark may not be as effective an analgesic as aspirin or NSAIDs, but supplementation may provide moderate pain relief according to a number of studies assessing patients with lower back pain and osteoarthritis [84, 91–93]. Despite being the only supplement listed for arthritis treatment by the FDA's report in 1999, conflicting results in the literature still exist. Biegart et al. found no significant osteoarthritis relief in two randomized double-blinded control trials with 240 mg salicin/day [94]. In contrast, Schmid et al. found significant pain relief with 240 mg salicin/day in a double-blind placebo-controlled trial over a 2-week period [93]. Based on the available evidence in the literature, willow bark should be considered a safe alternative to aspirin in the management of patient with lower back pain. However, these patients should be made aware that willow bark is considered to have "moderate analgesic effect in osteoarthritis" [93] and may not be as effective as NSAIDs or aspirin.

Conclusion

Nutraceuticals may be considered for patients with osteoarthritis who are not ready for a total joint replacement or are dissatisfied current management strategies. A summary of all nutraceuticals discussed is included in Table 7.1, and for patients with cost concerns, a daily cost estimate is provided in Table 7.2. Overall, most supplements are safe for use; however, none of the nutraceuticals discussed have strong supporting evidence of overall efficacy. Glucosamine/chondroitin, omega-3 fatty acids, ASUs, and willow bark can be tried for potential short-term relief with relatively few side effects. Vitamins and minerals should be considered with extreme caution due to potential side effects and lack of strong supporting evidence. Physician or nutritionist consultation for any patient is crucial for overall safety assurance with any nutraceutical. More high-level studies are needed before strong recommendations can be made about overall utility in osteoarthritis development and progression.

Table 7.1 Summary of nutraceuticals discussed in the chapter and patient considerations with supplementations

Nutraceutical name	Possible mechanism of action	Recommended dose for women[a]	Precautions[a]	Drug interactions w(minor (L), moderate (M), major (S))[a]	Toxicity side effects[a]	Other uses[a]
Glucosamine	Enhance chondrocyte synthesis by stimulating proteoglycan production, reduce proteolytic effects on chondrocytes and reduce expression of inflammatory factors	500 mg orally 3 times per day	Allergy to shellfish	Etoposide (S), teniposide (S), doxorubicin (S), warfarin (M)	Nausea, vomiting, diarrhea, abdominal pain	Few other uses other than osteoarthritis
Chondroitin	Enhance chondrocyte synthesis by stimulating proteoglycan production, reduce proteolytic effects on chondrocytes and reduce expression of inflammatory factors	800–1,200 mg daily (single or divided dose)	Unknown	Anticoagulants (warfarin and coumadin) (M)	Nausea, vomiting, diarrhea, abdominal pain	Corneal transplant, cataract surgery preservation, osteoarthritis, snoring, TMJ, xerophthalmia
ASUs	Inhibit inflammatory cytokines, stimulate collagen and aggrecan synthesis, and decrease spontaneous collagenase activity	300 mg daily	Allergy to latex	MAOI (S), warfarin (M),	Food intolerance	Hypercholesterolemia
Omega-3 Fatty Acids	Role in COX and LOX anti-inflammatory enzyme pathway	2,800 mg daily (40 mg/kg body weight)	Pregnant and lactating women, type 2 diabetes, fish or shellfish allergy	Anticoagulants (M)	Burping, indigestion, atrial fibrillation, vomiting, increased liver function, anaphylaxis	Hyperlipidemia, coronary arteriosclerosis

(continued)

Table 7.1 (continued)

Nutraceutical name	Possible mechanism of action	Recommended dose for women[a]	Precautions[a]	Drug interactions (minor (L), moderate (M), major (S))[a]	Toxicity side effects[a]	Other uses[a]
Vitamin A	Antioxidant properties	700 mcg/day	Pregnant and lactating women, hypervitaminosis A	Bexarotene (M), minocycline (M), tretinoin (M), etretinate (M), acitretin (M), isotretinoin (M), colestipol (L), carob (L)	Vomiting, anorexia, fatigue, irritability, skin lesions, increased intracranial pressure, radiograph periosteal calcification, yellow skin discoloration	Wound healing, low birth weight, vitamin A deficiency
Vitamin C	Antioxidant properties	90 mg/day for women (100–250 mg for deficiency)	Concurrent anticoagulant therapy, concurrent sodium restricted diet, diabetes, pregnancy	Aluminum-containing compounds, amygdalin, cyanocobalamin, deferoxamine, aminoacetate, sodium carbonate, indinavir, magaldrate	Diarrhea, iron overload, nephrolithiasis (high doses), hyperglycemia	Wound healing
Vitamin E	Antioxidant properties	15 mg/day (RDA)	Lactating women, facial chemical peel, blood clotting disorder	Dicumarol (S), warfarin (M), colestipol (L), orlistat (L)	Heart failure (long term), bleeding, sepsis, pulmonary embolism	Vitamin E deficiency
Boron	Alter other mineral levels in the body (i.e., calcium or magnesium); alter hormone levels (i.e., estrogen); and alter reactive oxygen species presence	6 mg daily	Phosphorus deficiency	Magnesium and phosphorus supplements	Riboflavonuria, tremors, convulsions, weakness, dermatitis, and hematemesis	Boron deficiency, diabetes

Zinc	Unknown	25–50 mg/day	Copper deficiency, glaucoma, homozygosity for hemochromatosis	Eltrombopag (S), iron (M), grepafloxacin (M), ofloxacin (M), moxifloxacin (M), tetracycline (M), gatifloxacin (M), sparfloxacin (M), gemifloxacin (M), norfloxacin (M), enoxacin (M), ciprofloxacin (M), levofloxacin (M), cinoxacin (M), penicillamine (L), copper (L), dairy food (L), caffeine (L)	Nausea, vomiting, and diarrhea	Wilson's disease, common cold, wound healing, zinc deficiency
Copper	Unknown	2–5 mg/day (always check blood levels prior to administration)	Wilson's disease	Zinc, molybdenum	Neutropenia/anemia, impaired coordination, death, epigastric burning, hypotension, jaundice, seizures, coma, shock, and death.	Copper deficiency, Menkes disease
Willow bark (salicin)	Analgesic	3 g/day divided in doses (ex. 325–650 mg Q4)	Aspirin allergy; asthma, alcohol use, pregnancy (third trimester), renal failure	Ketorolac (S), influenza virus vaccine live (S), varicella vaccine (S), ginko (S), anticoagulants (possibly), captopril (M), streptokinase (M) (selected)	Gastrointestinal ulcer, bronchospasm, Reye's syndrome, angioedema, tinnitus	Antipyretic, analgesic, antiphlogistic[b]

Note: All precautions, drug interactions, and side effects are selected and not comprehensive

[a]Recommended dose, precautions, drug interactions, toxicity side effects, and other uses from Thompson's *Micromedex*. http://www.micromeded.com. Accessed Feb 2013

[b]Blumenthal M, Busse WR, Goldberg A, et al. Herb guide by pharmacologic action (approved herbs). In: The complete German Commission E monographs: therapeutic guide to herbal medicines. Austin: The American Botanical Council; 1998. p. 462–69

Table 7.2 Cost comparison for daily nutraceutical use

Nutraceutical name	Cost comparison (per day)[a]
Glucosamine	$.33 combined with chondroitin
Chondroitin	$.33 combined with glucosamine
ASU	$.69
Fish oil	$.19 (omega-3 EPA and DHA)
Vitamin A	$.03
Vitamin C	$.27
Vitamin E	$.07
Zinc	$.04
Boron	$.08
Copper	$.05
Willow bark	$.08

[a]Cost data from http://vitacost.com. Accessed Jan 2013

References

1. Dietary Supplement Health and Education Act of 1994. Edited by Congress, Pub. L.; 1994. www.fda.gov/Food/DietarySupplements/default.htm
2. Cohen PA. Accessing supplement safety- the FDA's controversial proposal. N Engl J Med. 2012;366(5):389–91.
3. Hoffman JM, Gambhir SS, Kelloff GJ. Regulatory and reimbursement challenges for molecular imaging. Radiology. 2007;245(3):645–60.
4. Gunther S, Patterson RE, Kristal AR, Stratton KL, White E. Demographic and health-related correlates of herbal and specialty supplement use. J Am Diet Assoc. 2004;104(1):27–34.
5. Mintel International Limited. Complementary Medicines Market Intelligence. London: Mintel International Ltd; 2003.
6. Kalra K. Nutraceutical-definition and introduction. AAPS Pharm Sci. 2003;5(3):27–8.
7. Ziesel SH. Regulation of "Nutraceuticals". Science. 1999;285(5435):275–96.
8. Dahmer S, Schiller RM. Glucosamine. Am Fam Physician. 2008;78(4):471–6.
9. Jersoch J. Effects of glucosamine and chondroitin sulfate on cartilage metabolism in OA: outlook on other nutrient partners especially omega-3 fatty acids. Int J Rheumatol. 2011;969012: 1–17.
10. Black C, Clar C, Henderson R, et al. The clinical effectiveness of glucosamine and chondroitin supplements in slowing or arresting progression of osteoarthritis of the knee: a systematic review and economic evaluation. Health Technol Access. 2009;13(52):1–148.
11. Sakai S, Otake E, Toida T, Goda Y. Identification of the origin of chondroitin sulfate in "health foods". Chem Pharm Bull (Tokyo). 2007;55(2):299–303.
12. Jawed H, Anjum S, Awan SI, Simjee SU. Anti-arthritic effect of GN1, a novel synthetic analog of glucosamine, in the collagen-induced arthritis model in rats. Inflamm Res. 2011;60(12):1113–20.
13. Bassleer C, Rovati L, Franchimont P. Stimulation of proteoglycan production by glucosamine sulfate in chondrocytes isolated from human osteoarthritic articular cartilage in vitro. Osteoarthritis Cartilage. 1998;6(6):427–34.
14. Jerosch J. Effects of glucosamine and chondroitin sulfate on cartilage metabolism in OA: outlook on other nutrient partners especially omega-3 fatty acids. Int J Rheumatol. 2011;2011: 969012.
15. Wang SX, Laverty S, Dumitriu M, Plaas A, Grynpas MD. The effects of glucosamine hydrochloride on subchondral bone changes in an animal model of osteoarthritis. Arthritis Rheum. 2007;56(5):1537–48.
16. Clegg DO, et al. Glucosamine, chondroitin sulfate, and the two in combination for painful knee osteoarthritis. N Engl J Med. 2006;354(8):795–808.

17. Martin MS, VanSell S, Danter J. Glucosamine and chondroitin: an appropriate adjunct treatment of symptomatic osteoarthritis of the knee. Orthop Nurs. 2012;31(3):160–6.
18. Petersen SG, Saxne T, Heingard D, Holm L, Koskinen S, Stordal C, Christensen H, Aagaard P, Kjaer M. Glucosamine but not ibuprofen alters cartilage turnover in osteoarthritis patients in response to physical training. Osteoarthritis Cartilage. 2010;18(1):34–40.
19. Woodhouse LJ. Do Ibuprofen or glucosamine in addition to resistance exercise training improve muscle strength in patients with knee osteoarthritis? Clin J Sport Med. 2012;22(3):290–2.
20. Petersen SG, Beyer N, Hansen M, Holm L, Aagaard P, Mackey AL, Kjaer M. Nonsteroidal anti-inflammatory drug or glucosamine reduced pain and improved muscle strength with resistance training in a randomized controlled trial of knee osteoarthritis patients. Arch Phys Med Rehabil. 2011;92(8):1185–93.
21. Magrans-Courtney T, et al. Effects of diet type and supplementation of glucosamine, chondroitin, and MSM on body composition, functional status, and markers of health in women with knee osteoarthritis initiating a resistance-based exercise and weight loss program. J Int Soc Sports Nutr. 2011;8(1):8.
22. Muller-Fassbender H, Bach GL, Haase W, Rovati LC, Setnikar I. Glucosamine sulfate compared to ibuprofen in osteoarthritis of the knee. Osteoarthritis Cartilage. 1994;2(1):61–9.
23. Lopes Vaz A. Double-blind clinical evaluation of the relative efficacy of ibuprofen and glucosamine sulphate in the management of osteoarthrosis of the knee in out-patients. Curr Med Res Opin. 1982;8(3):145–9.
24. Morreale P, Manopulo R, Galati M, et al. Comparison of the anti-inflammatory efficacy of chondroitin sulfate and diclofenac sodium in patients with knee osteoarthritis. J Rheumatol. 1996;23:1385–91.
25. Craig DG, Ford AC, Hayes PC, Simpson KJ. Systematic review: prognostic tests of paracetamol-induced acute liver failure. Aliment Pharmacol Ther. 2010;31(10):1064–76.
26. Larson AM, et al. Acetaminophen-induced acute liver failure: results of a United States multicenter, prospective study. Hepatology. 2005;42(6):1364–72.
27. Fact Sheet- Osteoarthritis. Edited, National Institutes of Health; 2010. report.nih.gov/nihfactsheets/ViewFactSheet.aspx?csid=55
28. Herrero-Beaumont G, et al. Glucosamine sulfate in the treatment of knee osteoarthritis symptoms: a randomized, double-blind, placebo-controlled study using acetaminophen as a side comparator. Arthritis Rheum. 2007;56(2):555–67.
29. Scroggie DA, Albright A, Harris MD. The effect of glucosamine-chondroitin supplementation on glycosylated hemoglobin levels in patients with type 2 diabetes mellitus: a placebo-controlled, double-blinded, randomized clinical trial. Arch Intern Med. 2003;163(13):1587–90.
30. Tannis AJ, Barban J, Conquer JA. Effect of glucosamine supplementation on fasting and non-fasting plasma glucose and serum insulin concentrations in healthy individuals. Osteoarthritis Cartilage. 2004;12(6):506–11.
31. Sherman AL, Ojeda-Correal G, Mena J. Use of glucosamine and chondroitin in persons with osteoarthritis. PM&R. 2012;4:S110–6.
32. Pavelka K, Gatterova J, Olejarova M. Glucosamine sulfate use and delay of progression of knee osteoarthritis: a 3-year, randomized, placebo-controlled, double blind study. Arch Intern Med. 2002;162:2113–23.
33. Michel BA, Stucki G, Frey D, De Vathaire F, Vignon E, Bruehlmann P, Uebelhart D. Chondroitins 4 and 6 sulfate in osteoarthritis of the knee: a randomized, controlled trial. Arthritis Rheum. 2005;52(3):779–86.
34. Jackson CG, Plaas AH, Sandy JD, Hua C, Kim-Rolands S, Barnhill JG, Harris CL, Clegg DO. The human pharmacokinetics of oral ingestion of glucosamine and chondroitin sulfate taken separately or in combination. Osteoarthritis Cartilage. 2010;18(3):297–302.
35. Richmond J, Hunter D, Irrgang J, et al. Clinical practice guideline on the treatment of osteoarthritis of the knee (non-arthroplasty). Rosemont: American Academy of Orthopedic Surgeons; 2008.
36. Simopoulos AP. The importance of the omega-6/omega-3 fatty acid ratio in cardiovascular disease and other chronic diseases. Exp Biol Med. 2008;233:674–88.
37. Kashiyama T, Ueda Y, Nemoto T, et al. Relationship between coronary plaque vulnerability and serum n-3/n-6 polyunsaturated fatty acid ratio. Circ J. 2011;75:2432–8.

38. Kremer JM, Jubiz W, Michalek A, Rynes RI, Bartholomew LE, Bigaouette J, Timchalk M, Beeler D, Lininger L. Fish-oil fatty acid supplementation in active rheumatoid arthritis. A double-blinded, controlled, crossover study. Ann Intern Med. 1987;106(4):497–503.
39. Galarraga B, Ho M, Youssef HM, Hill A, McMahon H, Hall C, Ogston S, Nuki G, Belch JJ. Cod liver oil (n-3 fatty acids) as an non-steroidal anti-inflammatory drug sparing agent in rheumatoid arthritis. Rheumatology (Oxford). 2008;47(5):665–9.
40. Lopez HL. Nutritional interventions to prevent and treat osteoarthritis. Part I: focus on fatty acids and macronutrients. PM&R. 2012;4:S145–54.
41. Baker KR, et al. Association of plasma n-6 and n-3 polyunsaturated fatty acids with synovitis in the knee: the MOST study. Osteoarthritis Cartilage. 2012;20(5):382–7.
42. Stammers T, Sibbald B, Freeling P. Fish oil in osteoarthritis. Lancet. 1989;334(8661):503.
43. Stammers T, Sibbald B, Freeling P. Efficacy of cod liver oil as an adjunct to non-steroidal anti-inflammatory drug treatment in the management of osteoarthritis in general practice. Ann Rheum Dis. 1992;27(10):128–9.
44. Deutsch L. Evaluation of the effect of Neptune Krill Oil on chronic inflammation and arthritic symptoms. J Am Coll Nutr. 2007;26:39–48.
45. Mauviel A, Daireaux M, Hartmann DJ, Galera P, Loyau G, Pujol JP. Effects of unsaponifiable extracts of avocado/soy beans (PIAS) on the production of collagen by cultures of synoviocytes, articular chondrocytes and skin fibroblasts. Rev Rhum Mal Osteoartic. 1989;56(2): 207–11.
46. Henrotin YE, Sanchez C, Deberg MA, Piccardi N, Guillou GB, Msika P, Reginster JY. Avocado/soybean unsaponifiables increase aggrecan synthesis and reduce catabolic and proinflammatory mediator production by human osteoarthritic chondrocytes. J Rheumatol. 2003; 30(8):1825–34.
47. Msika P, Baudouin C, Saunois A, Bauer T. Avocado/soybean unsaponifiables, ASU EXPANSCIENCE, are strictly different from the nutraceutical products claiming ASU appellation. Osteoarthritis Cartilage. 2008;16(10):1275–6.
48. Appelboom T, Schuermans J, Verbruggen G, Henrotin Y, Reginster JY. Symptoms modifying effect of avocado/soybean unsaponifiables (ASU) in knee osteoarthritis. A double blind, prospective, placebo-controlled study. Scand J Rheumatol. 2001;30(4):242–7.
49. Maheu E, Mazieres B, Valat JP, Loyau G, Le Loet X, Bourgeois P, Grouin JM, Rozenberg S. Symptomatic efficacy of avocado/soybean unsaponifiables in the treatment of osteoarthritis of the knee and hip: a prospective, randomized, double-blind, placebo-controlled, multicenter clinical trial with a six-month treatment period and a two-month followup demonstrating a persistent effect. Arthritis Rheum. 1998;41(1):81–91.
50. Blotman F, Maheu E, Wulwik A, Caspard H, Lopez A. Efficacy and safety of avocado/soybean unsaponifiables in the treatment of symptomatic osteoarthritis of the knee and hip. A prospective, multicenter, three-month, randomized, double-blind, placebo-controlled trial. Rev Rhum Engl Ed. 1997;64(12):825–34.
51. Maheu E, Cadet C, Marty M, et al. Evaluation of the structure modifying effects of the avocado-soybean unsaponifiables (ASU): results of the ERADUAS study, a 3-year prospective, randomized, double-blind placebo controlled trial. Osteoarthritis Cartilage. 2010;18(S2): S45–256.
52. Lequesne M, Maheu E, Cadet C, Dreiser RL. Structural effect of avocado/soybean unsaponifiables on joint space loss in osteoarthritis of the hip. Arthritis Rheum. 2002;47(1):50–8.
53. Walker-Bone K. "Natural remedies" in the treatment of osteoarthritis. Drugs Aging. 2003; 20(7):517–26.
54. Christensen R, Bartels EM, Astrup A, Bliddal H. Symptomatic efficacy of avocado-soybean unsaponifiables (ASU) in osteoarthritis (OA) patients: a meta-analysis of randomized controlled trials. Osteoarthritis Cartilage. 2008;16(4):399–408.
55. Olivier P, Montastruc JL. Post-marketing safety profile of avocado-soybean unsaponifiables. Presse Med. 2010;39(10):e211–6.

56. Canter PH, Wider B, Ernst E. The antioxidant vitamins A, C, E and selenium in the treatment of arthritis: a systematic review of randomized clinical trials. Rheumatology (Oxford). 2007;46(8):1223–33.
57. Mahmud Z, Ali SM. Role of vitamin A and E in spondylosis. Bangladesh Med Res Counc Bull. 1992;18(1):47–59.
58. De Roos AJ, Arab L, Renner JB, Craft N, Luta G, Helmick CG, Hochberg MC, Jordan JM. Serum carotenoids and radiographic knee osteoarthritis: the Johnston County Osteoarthritis Project. Public Health Nutr. 2001;4(5):935–42.
59. McAlindon T, Jacques P, Zhang Y, Hannan MT, Alibadi P, Weissman B, Rush D, Levy D, Felson DT. Do antioxidant micronutrients protect against the development and progression of knee osteoarthritis? Arthritis Rheum. 1996;39:648–56.
60. Vitamin A. In: MICROMEDEX [database on the Internet]. 2013. Available from: www.thomsonhc.com/micromedex2. Subscription required to view. Updated 8 Feb 2013; cited 12 Feb 2013.
61. Dugdale D, Zieve D. Hypervitaminosis A. In: A.D.A.M Medical Encyclopedia. Edited, PubMed Health, A.D.A.M., Inc., Bethesda, MD. 2012.
62. Kurz B, Jost B, Schunke M. Dietary vitamins and selenium diminish the development of mechanically induced osteoarthritis and increase the expression of antioxidative enzymes in the knee joint of STR/1N mice. Osteoarthritis Cartilage. 2002;10(2):119–26.
63. Sowers M, Lachance L. Vitamins and arthritis. The roles of vitamins A, C, D, and E. Rheum Dis Clin North Am. 1999;25(2):315–32.
64. McAlindon T, Bigee B. Nutritional factors and osteoarthritis: recent developments. Curr Opin Rheumatol. 2005;17:647–52.
65. Jensen NH. Reduced pain from osteoarthritis in hip joint or knee joint during treatment with calcium ascorbate. A randomized, placebo-controlled cross-over trial in general practice. Ugeskr Laeger. 2003;165(25):2563–6.
66. Peregoy J, Wilder FV. The effects of vitamin C supplementation on incident and progressive knee osteoarthritis: a longitudinal study. Public Health Nutr. 2011;14(4):709–15.
67. Bhattacharya I, Saxena R, Gupta V. Efficacy of vitamin E in knee osteoarthritis management of North Indian geriatric population. Ther Adv Musculoskelet Dis. 2012;4(1):11–9.
68. Wluka A, Stuckey S, Brand C, Cicuttini F. Supplementary vitamin E does not affect the loss of cartilage volume in knee osteoarthritis: a 2 year double blind randomized placebo controlled study. J Rheumatol. 2002;29(12):2585–91.
69. Miller 3rd ER, Pastor-Barriuso R, Dalal D, Riemersma RA, Appel LJ, Guallar E. Meta-analysis: high-dosage vitamin E supplementation may increase all-cause mortality. Ann Intern Med. 2005;142(1):37–46.
70. Hunt CD, Shuler TR, Mullen LM. Concentration of boron and other elements in human foods and personal-care products. J Am Diet Assoc. 1991;91(5):558–68.
71. Nielsen F. The emergence of boron as nutritionally important throughout the life cycle. Nutrition. 2000;16(7–8):512–4.
72. Nielsen F. The justification for providing dietary guidance for the nutritional intake of boron. Biol Trace Elem Res. 1998;66(1–3):319–30.
73. Newnham R. Mineral imbalance and boron deficiency. In: Trace element metabolism in man and animals (TEMA-4). Australian Acad. Sci. Canberra. 1981. p. 400–2.
74. Newnham RE. Essentiality of boron for healthy bones and joints. Environ Health Perspect. 1994;102 Suppl 7:83–5.
75. Hakki SS, Dundar N, Kayis SA, Hakki EE, Hamurcu M, Kerimoglu U, Baspinar N, Basoglu A, Nielsen FH. Boron enhances strength and alters mineral composition of bone in rabbits fed a high energy diet. J Trace Elem Med Biol. 2013;27(2):148–53.
76. Meacham S, Taper J, Volpoe S. Effects of boron supplementation on bone mineral density and dietary, blood and urinary calcium, phosphorus, magnesium, and boron in female athletes. Environ Health Perspect. 1994;102(S7):79–82.
77. Fail PA, Chapin RE, Price CJ, Heindel JJ. General, reproductive, developmental, and endocrine toxicity of boronated compounds. Reprod Toxicol. 1998;12(1):1–18.

78. Anon: Boron. In: DerMarderosian A (ed.). The Lawrence Review of Natural Products. Facts and Comparisons, St. Louis, MO; 1995.
79. Nielsen F. Is boron nutritionally relevant? Nutr Rev. 2008;66(4):183–91.
80. Nielsen FH. Biochemical and physiologic consequences of boron deprivation in humans. Environ Health Perspect. 1994;102 Suppl 7:59–63.
81. Grennan DM, Knudson JM, Dunckley J, MacKinnon MJ, Myers DB, Palmer DG. Serum copper and zinc in rheumatoid arthritis and osteoarthritis. N Z Med J. 1980;91(652):47–50.
82. Yazar M, Sarban S, Kocyigit A, Isikan UE. Synovial fluid and plasma selenium, copper, zinc, and iron concentrations in patients with rheumatoid arthritis and osteoarthritis. Biol Trace Elem Res. 2005;106(2):123–32.
83. Zinc. In: MicroMedex. [database on the Internet]. 2012. Available from: www.thomsonhc.com/micromedex2. Subscription required to view. Updated 6 Dec 2012; cited 12 Feb 2013.
84. Salix. In MicroMedex. [database on the Internet]. 2012. Available from: www.thomsonhc.com/micromedex2. Subscription required to view. Updated 2012; cited 12 Feb 2013.
85. Norn S, Permin H, Kruse PR, Kruse E. From willow bark to acetylsalicylic acid. Dan Medicinhist Arbog. 2009;37:79–98.
86. Krivoy N, Pavlotzky E, Chrubasik S, Eisenberg E, Brook G. Effect of salicis cortex extract on human platelet aggregation. Planta Med. 2001;67(3):209–12.
87. Marz RW, Kemper F. Willow bark extract-effects and effectiveness. Status of current knowledge regarding pharmacology, toxicology and clinical aspects. Wien Med Wochenschr. 2002;152(15–16):354–9.
88. Boullata JI, McDonnell PJ, Oliva CD. Anaphylactic reaction to a dietary supplement containing willow bark. Ann Pharmacother. 2003;37(6):832–5.
89. Jennings A. Link between salicylate and willow bark. Pharm J. 2006;276:417.
90. Setty AR, Sigal LH. Herbal medications commonly used in the practice of rheumatology: mechanisms of action, efficacy, and side effects. Semin Arthritis Rheum. 2005;34(6):773–84.
91. Chrubasik S, Kunzel O, Model A, Conradt C, Black A. Treatment of low back pain with a herbal or synthetic anti-rheumatic: a randomized controlled study. Willow bark extract for low back pain. Rheumatology (Oxford). 2001;40(12):1388–93.
92. Chrubasik S, Eisenberg E, Balan E, Weinberger T, Luzzati R, Conradt C. Treatment of low back pain exacerbations with willow bark extract: a randomized double-blind study. Am J Med. 2000;109(1):9–14.
93. Schmid B, Ludtke R, Selbmann HK, Kotter I, Tschirdewahn B, Schaffner W, Heide L. Efficacy and tolerability of a standardized willow bark extract in patients with osteoarthritis: randomized placebo-controlled, double blind clinical trial. Phytother Res. 2001;15(4):344–50.
94. Biegert C, Wagner I, Ludtke R, Kotter I, Lohmuller C, Gunaydin I, Taxis K, Heide L. Efficacy and safety of willow bark extract in the treatment of osteoarthritis and rheumatoid arthritis: results of 2 randomized double-blind controlled trials. J Rheumatol. 2004;31(11):2121–30.
95. Nelson DL, Cox MM. Lehninger principles of biochemistry. New York: WH Freeman & Company; 2009.

Chapter 8
Alternative Exercise for Women

Elinor Mody

Abstract There is scientific evidence that women are more prone to specific injuries than are men, due to differences in hip and knee anatomy and hormonal profiles. Aging women are also susceptible to the "Dowager's Hump," or kyphosis, and decreased balance. Various factors are responsible for these conditions, including hormones and joint angles. Currently popular exercise programs, yoga and Pilates, have been shown to have some positive impact on these conditions. This chapter explores the nature of these conditions and the impact of yoga and Pilates.

Keywords Women • Musculoskeletal • Yoga • Pilates • Kyphosis

Introduction

Scientific data to support the fact that women are prone to different musculoskeletal injuries and syndromes than men is growing rapidly and has been for the past several years. Gender is also an important factor in sports injuries. In particular, there is evidence that women are more prone to specific injuries than are men, due to differences in hip and knee anatomy and hormonal profiles. Aging women are also susceptible to the "Dowager's Hump," or kyphosis, and decreased balance. Data are also amassing that some of these tendencies can be lessened by exercise. In this chapter, these musculoskeletal problems will be explored, as will the data to support two popular types of exercise, yoga and Pilates.

The best-known injury that most commonly affects women is the anterior cruciate ligament (ACL) tear. Depending on the sport, this injury is up to four times more

E. Mody, MD
Department of Rheumatology, Gretchen and Edward
Fish Center for Women's Health, Brigham and Women's Hospital,
850 Boyleston Street, Suite 402, Chestnut Hill, MA 02467, USA
e-mail: emody@partners.org

E. Mody, E. Matzkin (eds.), *Musculoskeletal Health in Women*,
DOI 10.1007/978-1-4471-4712-1_8, © Springer-Verlag London 2014

common in women than men [1]. There are various potential reasons for this, including a reduced femoral intercondylar notch width, increased ligamentous laxity due to higher estrogen levels [2], and valgus torque due to hip and knee biomechanics [3]. Women are also known to have a higher incidence of ankle sprain than men [4].

There may be reasons beyond anatomical and physiological differences that are responsible for the increased injury rate in women. In particular, it has been suggested women have different hand-eye coordination skills than men [5], which can lead to an increase in injuries, due in part to decreased balance. A recent article suggested that gender differences in motor development may be apparent in childhood [6]. Additionally, men and women exhibit different responses to exercise [7]. Strength and conditioning programs should be modified to account for these gender differences. For example, gender differences in studies on jumping have demonstrated that women benefit from increased strength and power at the hip, knee, and ankle in order to jump as well as men [8].

What does all of this mean to us? How do we use this information to better help our patients? While there are some studies in the literature, this is an under-researched area of medicine. Since menstrual cycle and femoral notch shape are not modifiable risk factors for ACL injury, research has focused on correcting neuromuscular imbalances in women to prevent these injuries. For example, increased knee abduction and imbalance have been found to be predictors of ACL injury [9]. Muscle strengthening with special emphasis on the terminal knee extensors and hip abductors can improve these neuromuscular imbalances. As physicians, we must try to educate our female patients about this data. We also need to have at our disposal concrete suggestions about the types of exercise to recommend. The data available in the literature to support the use of two currently popular forms of exercise are reviewed. In particular, the literature supporting the use of yoga and Pilates for spinal flexibility, kyphosis, balance, and muscle strengthening is examined. Additionally, taping and bracing at the ankle may be helpful, particularly in women prone to certain injuries.

Pilates

The Pilates method is named after Josef Pilates, who developed a system of exercise in the early 1900s, which was designed to rehabilitate injured ballet dancers. The method incorporates "core" muscle strengthening, including the abdominal and mid/lower back muscles. This is done by engaging in "destabilizing" exercises, which requires muscle activation to stabilize the body. These exercises also help with balance training as a result of destabilization, which is very important for the aging female. Additionally, the Pilates method also incorporates spinal flexion and extension. In theory, this method may help with kyphosis and low back pain that are common in older women. These exercises are either practiced on the mat or on various pieces of equipment: the reformer, the Cadillac, the chair (Fig. 8.1), and the barrel (Fig. 8.2). The equipment provides support and aids in the practice of Pilates.

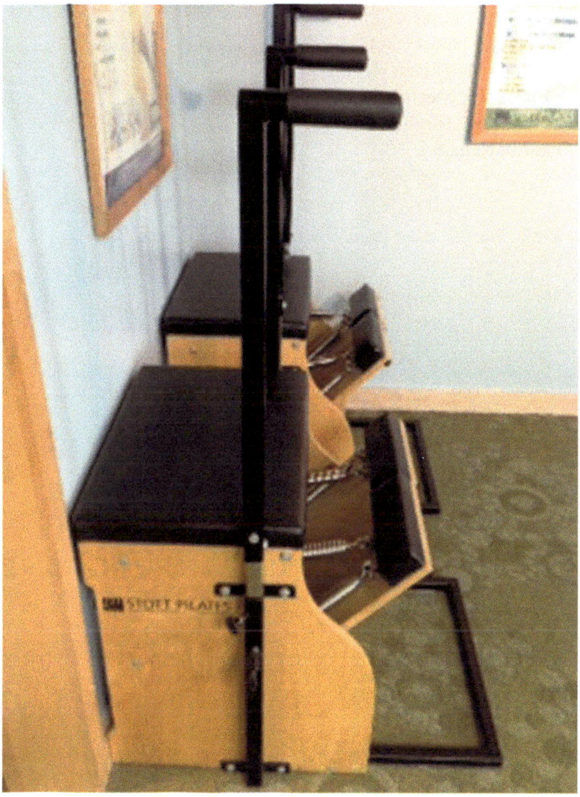

Fig. 8.1 Example Pilates equipment: chair (Stott Pilates®, Merrithew Health & Fitness™, Merrithew Corporation, Toronto, Ontario, Canada)

This section explores the use of Pilates for spinal pain and lack of extension, in addition to aiding in proprioception and balance.

Spinal Pain Disorders

There are many small studies investigating the use of Pilates in patients with chronic low back pain (LBP). Pilates is perceived as useful by patients, as evidenced by a survey performed in patients with LBP taking a Pilates class in the UK [10]. In a study by Dolan et al., 180 patients were referred by the same neurosurgeon for chronic LBP to three different physical therapy practices. Patients who completed the course had a significantly better outcome measured by the Low Back Outcome Score, compared to patients that did not complete the course [11]. Pilates has also been compared in multiple studies to placebo, and general low back stabilization. A meta-analysis done by Pereira et al. looks at five studies of this nature. Although Pilates was not shown to be significantly better than general lower back stabilization exercises in terms of functionality, pain relief was better than placebo. However,

Fig. 8.2 Example of Pilates equipment: barrel (Stott Pilates®, Merrithew Health & Fitness™, Merrithew Corporation, Toronto, Ontario, Canada)

follow-up was limited to a maximum of 8 weeks [12]. In a randomized controlled trial done by Brooks et al., Pilates exercises were compared to stationary bike exercise in patients with chronic low back pain for a period of 8 weeks. There was no confounding variable in the control group, such as core strengthening in the control group, which was cited as a problem in other studies. The patients randomized to the Pilates arm showed a significant decrease in self-rated disability and self-rated pain. Additionally, this study incorporated the use of EMG to determine the anticipatory postural adjustment (APA), so the authors were able to show that there was a significant decrease in delayed APAs (commonly seen in patients with chronic low back pain) in the Pilates group [13].

Balance/Flexibility

Studies have examined the effect of Pilates on balance. In a study done by Hall et al., patients were randomized to a 10-week course of Pilates consisting of traditional strength and flexibility training or no exercise (control arm). Pre- and post-training measurements of static and dynamic balance were done on the Kinesthetic

Ability Training balance platform. Static balance was significantly improved in the Pilates group; all three groups improved on dynamic balance measurement [14]. Pilates has also been shown to improve thoracic kyphosis in a group of subjects over the age of 60. However, these patients did not suffer from chronic pain, and these patients were older than the typical Pilates exerciser [15]. Pilates has also been studied in increasing general spinal flexibility. In a study by Schroeder et al., young, healthy men underwent three Pilates reformer sessions. Various measures of flexibility were used immediately before and after each session. A significant increase in spine flexibility was noted [16].

Although most of these studies are quite small, in sum, they support the use of Pilates in various situations. As discussed previously, women are more prone than men to certain types of injuries, kyphosis, and a decrease in static and dynamic balance. Additionally, women are more prone to osteoporosis as they age than are men, making balance an even more important ability, as lack of balance can lead to osteoporotic fractures. Therefore, although more data is needed, the practice of Pilates appears to be an important potential choice for mode of exercise in women.

Yoga

Yoga is an ancient practice and encompasses benefits for many organs of the body, including the lungs, the heart, the psyche, and of course, the musculoskeletal system [17]. Yoga has been shown to strengthen muscles, including the quadriceps, and has also been shown to improve balance and increase spinal flexibility [18]. All of these are critical for the musculoskeletal health of all women. In this chapter, the evidence in the medical literature to support the use of yoga will be explored.

Different Forms of Yoga

There are several different "flavors" of yoga. All of the forms of yoga practiced routinely in the USA are forms of Hatha yoga. The most commonly available yoga is Vinyasa flow. This is an athletic practice of yoga, very fast moving, and often done in a warm room. There are various "slower" forms of yoga, including Kripalu, and Yin, or restorative. Regardless of the type of yoga practiced, if done correctly, then there will be benefits as mentioned previously.

Muscle Strengthening/Treatment of Injury

Most of the trials to evaluate yoga as a treatment for arthritis have been small, short, and not blinded. However, this is acceptable since the risk of trying yoga is minimal. There are some data to suggest that compliance is higher for yoga than other types of exercise [19].

Several yoga *asanas* used in the treatment of OA of the knee, which emphasize fully extending the knee, clearly can strengthen the quadriceps. Quadriceps strengthening is an important part of most approaches to treating knee OA. A case series by Bukowski et al. examines the effect of Iyengar yoga in patients with osteoarthritis of the knee. A group of 15 patients with osteoarthritis of the knee was randomized to one of three groups: yoga, traditional stretching and strengthening, or no structured group exercise. The WOMAC scale was used to assess subjective change. After 6 weeks, both the yoga group and the traditional exercise group were found to have functional and quality of life improvement [20].

There are many studies that evaluate the effect of yoga practice on low back pain, particularly chronic low back pain. In a randomized controlled trial done of intensive short-term yoga therapy in a group of patients with chronic low back pain, Tekur et al. found that after 7 days, the yoga group had a significant reduction in disability as measured by the Oswestry Disability Index (ODI) and also had an increase in spinal flexibility measured by goniometry [21]. In another trial by Sherman et al., 12 weekly yoga classes were compared to conventional stretching exercises and to a self-care program in 228 patients with chronic low back pain. The yoga group, compared to the conventional stretching group, was found to show significant improvement in functionality and pain as measured by the Roland-Morris Disability Questionnaire (RMDQ) [22]. In a randomized trial published in the Annals of Internal Medicine, written by Tilbrook et al., 313 patients with chronic low back pain were randomized to a 12-week yoga program or to a back pain education booklet and "usual care," which was not well described. The yoga group had a significant decrease in the RMDQ compared with the control group, and this was maintained at 6 months [23]. A similar trial was performed by Williams et al. over 24 weeks, and again, the yoga group was found to have significant decreases in pain and increased functionality measured by the ODI, and this change was maintained at 6-month follow-up [24]. A small study by Groessl et al. looked at yoga treatment in men and women veterans. Fifty-three veterans with chronic low back pain were treated with 10 weeks of yoga therapy. The women subjects were found to have significant decreases in depression measured by the CESD-10 scale and significantly larger increases in energy and SF-12 mental health than the male subjects [25]. Yoga for the treatment of low back pain has also been examined from a cost perspective. In a trial by Chuang et al., yoga and usual therapy was compared to usual therapy alone over a 12-week period in patients with chronic low back pain in the UK, taking into account willingness to pay [26].

Other types of musculoskeletal injury have also been studied with respect to yoga. Michalsen et al. randomized 77 patients with chronic neck pain to either a 9-week course of Iyengar yoga or a self-care/exercise. The yoga group was found to have significant improvements in pain at rest, pain in motion, pain-related apprehension, disability, quality of life, and psychological outcomes [27]. Yoga has also been used to treat computer injuries. Telles et al. studied 291 professional computer users for handgrip strength, tapping speed, low back pain, and hamstring flexibility. They were then randomized to yoga or no intervention. At the end of 60 days, the yoga group showed a significant decrease in the frequency, intensity, and degree of interference in their work due to discomfort and also showed a significant increase in handgrip strength and right-hand tapping speed [28].

Balance/Spinal Flexibility

Ability to balance, as discussed earlier in this chapter, decreases with age, particularly in women. Balance training is an integral part of all types of yoga. Most of us are familiar with some common yoga poses such as "tree" pose and "dancer" pose, both aimed at improving balance. Interestingly enough, how one deals with the frustration of not being able to achieve these poses is part of the practice of yoga. Additionally, yoga practice focuses on spinal extension and torsion. In theory, this could be useful in the treatment of kyphosis, which plagues older women. In fact, a mathematical model has been developed to describe the effects of the sun salutation postures on specific joints [29].

There are multiple studies looking at the effect of yoga on balance and gait in young and old, women and men. Hart et al. looked at Bikram yoga as steadiness training in young adults as compared to control (no training) over an 8-week period. Interestingly, an improvement in knee extensor steadiness, suggesting quadriceps strengthening, was found in the yoga group. Balance time was also improved [30]. In a small study of women with various musculoskeletal complaints, a 4-week trial of yoga was evaluated; pre- and posttreatment values for static balance and gait were measured with a stabilometer and gait trainer. Post-study values showed improvement in both balance and gait; these changes were significant [31]. In a pilot study by Zettergren et al., an 8-week yoga program was evaluated in improving the postural control, mobility, and gait speed in community-living elders (mean age in the 80s). Significant improvement was seen in balance, measured by the Berg Balance Score, and in walking speed [32]. In a study by Gonscalves et al., 83 women 60 or older were entered into a 14-week Hatha yoga program vs. a control group. Flexibility was increased in all body areas studied in the yoga group except for the elbow. The yoga group also decreased the time in which it took them to stand from a chair, take off a T-shirt, and various other ADLs [33]. Chen et al. randomized elders to either a "silver yoga" program or control and measured flexibility, muscle power, balance, and agility. All of these characteristics were significantly increased in the yoga group compared to control [34]. Schmid et al. studied the effect of yoga on fear of falling, which is also a significant problem in the elderly, often leading to isolation. In this open trial, 14 elders (over 65), who had a self-professed fear of falling, were enrolled in a 12-week yoga trial. Fear of falling (FoF) was measured with the Illinois FoF measure, and balance was measured with the Berg Balance Scale. Upper and lower body flexibility was measured with the back scratch test and chair sit and reach test. Improvement was seen in all areas, though it is not clear that statistical significance was seen [35]. DiBenedetto et al. looked at the effect of an 8-week Hatha yoga program on peak hip extension, average anterior pelvic tilt, and stride length at walking speed in 23 patients aged 62–83. Significant improvement was seen in peak hip extension and stride length [36]. Yoga has also been evaluated in poststroke patients. Schmid et al. studied the effect of an 8-week yoga program in poststroke patients' ability to balance assessed by the Berg Balance Scale. This was significantly increased in the patients who completed the 8-week course. Their FoF was significantly decreased as well [37].

Yoga has also been studied for its effect on decreasing kyphosis. Greendale et al. studied a 6-month course of yoga in a group of men and women over the age of 60 with a kyphosis angle of 40° or greater. These patients were randomized into treatment and control groups. The yoga group experienced a 4.4 % improvement in kyphosis angle, which easily reached statistical significance over control [38].

There is a significant body of data to support the use of yoga in patients suffering from osteoarthritis of various joints, in particular the low back, kyphosis, and decreased balance. Additionally, it is suggested that women respond to yoga therapy better than do men. Clearly more work needs to be done in this area, but in the meantime, practitioners should feel confident in recommending yoga to these appropriate patients.

Conclusion

In sum, women are prone to various musculoskeletal issues that men are not to the same degree. As physicians, we must recognize this and look for solutions based on the medical literature available. Although the evidence to support yoga and Pilates practice is not robust, it is suggestive that these forms of exercise may be useful in addressing many of the unique musculoskeletal problems that women face.

References

1. Hewett TE, Ford KR, Myer GD. Anterior cruciate ligament injuries in female athletes: part 2, a meta-analysis of neuromuscular interventions aimed at injury prevention. Am J Sports Med. 2006;34(3):490–8.
2. Hewett TE, Zazulak BT, Myer GD. Effects of the menstrual cycle on anterior cruciate ligament injury risk: a systematic review. Am J Sports Med. 2007;35(4):659–68.
3. Renstrom P, Ljungqvist A, Arendt E, Beynnon B, Fukubayashi T, Garrett W, Georgoulis T, Hewett TE, Johnson R, Krosshaug T. Non-contact ACL injuries in female athletes: an International Olympic Committee current concepts statement. Br J Sports Med. 2008;42(6):394–412.
4. O'Connor K, Bragdon G, Baumhauer JF. Sexual dimorphism of the foot and ankle. Orthop Clin North Am. 2006;37(4):569.
5. Ruff RM, Parker SB. Gender-and age-specific changes in motor speed and eye-hand coordination in adults: normative values for the Finger Tapping and Grooved Pegboard Tests. Percept Mot Skills. 1993;76(3c):1219–30.
6. Gidley Larson JC, Mostofsky SH, Goldberg MC, Cutting LE, Denckla MB, Mahone EM. Effects of gender and age on motor exam in typically developing children. Dev Neuropsychol. 2007;32(1):543–62.

7. Sewright K, Hubal M, Kearns A, Holbrook M, Clarkson P. Sex differences in response to maximal eccentric exercise. Med Sci Sports Exerc. 2008;40(2):242.
8. Lawson BR, Thomas M, Stephens I, DeVoe DE, Raoul F, Resier I. Lower-extremity bilateral differences during step-close and no-step countermovement jumps with concern for gender. J Strength Cond Res. 2006;20(3):608–19.
9. Schnurrer-Luke Vrbanić T, Ravlić-Gulan J, Gulan G, Matovinović D. Balance index score as a predictive factor for lower sports results or anterior cruciate ligament knee injuries in Croatian female athletes–preliminary study. Coll Antropol. 2007;31(1):253–8.
10. Gracey J, Galvin J. A survey to determine the perceptions of pilates as a form of exercise and as a treatment for low back pain. J Bone Joint Surg Br. 2010;92(Suppl I):238.
11. Dolan A, Hutchinson M, Fraser R. The Pilates based exercise programme in the management of low back pain. J Bone Joint Surg Br. 2001;83.
12. Pereira LM, Obara K, Dias JM, Menacho MO, Guariglia DA, Schiavoni D, Pereira HM, Cardoso JR. Comparing the Pilates method with no exercise or lumbar stabilization for pain and functionality in patients with chronic low back pain: systematic review and meta analysis. Clin Rehabil. 2012;26(1):10–20.
13. Brooks C, Kennedy S, Marshall PW. Specific trunk and general exercise elicit similar changes in anticipatory postural adjustments in patients with chronic low back pain: a randomized controlled trial. Spine. 2012;37(25):E1543–50.
14. Johnson EG, Larsen A, Ozawa H, Wilson CA, Kennedy KL. The effects of Pilates-based exercise on dynamic balance in healthy adults. J Bodyw Mov Ther. 2007;11(3):238–42.
15. Kuo Y-L, Tully EA, Galea MP. Sagittal spinal posture after Pilates-based exercise in healthy older adults. Spine. 2009;34(10):1046–51.
16. Schroeder JM, Crussemeyer JA, Newton SJ. Flexibility and heart rate response to an acute Pilates reformer session. Med Sci Sports Exerc. 2002;34(5):S258.
17. Thombre D, Balakumar B, Nambinarayanan T, Thakur S, Krishnamurthy N, Chandrabose A. Effect of yoga training on reaction time, respiratory endurance and muscle strength. Indian J Physiol Pharmacol. 1992;36(4):229.
18. Sorosky S, Stilp S, Akuthota V. Yoga and pilates in the management of low back pain. Curr Rev Musculoskelet Med. 2008;1(1):39–47.
19. Haaz S, Bartlett SJ. Yoga for arthritis: a scoping review. Rheum Dis Clin North Am. 2011;37(1):33.
20. Bukowski EL, Conway A, Glentz LA, Kurland K, Galantino ML. The effect of iyengar yoga and strengthening exercises for people living with osteoarthritis of the knee: a case series. Int Q Community Health Educ. 2006;26(3):287–305.
21. Tekur P, Singphow C, Nagendra HR, Raghuram N. Effect of short-term intensive yoga program on pain, functional disability and spinal flexibility in chronic low back pain: a randomized control study. J Altern Complement Med. 2008;14(6):637–44.
22. Sherman KJ, Cherkin DC, Wellman RD, Cook AJ, Hawkes RJ, Delaney K, Deyo RA. A randomized trial comparing yoga, stretching, and a self-care book for chronic low back pain. Arch Intern Med. 2011;2524:v2011.
23. Tilbrook HE, Cox H, Hewitt CE, Kang'ombe AR, Chuang L-H, Jayakody S, Aplin JD, Semlyen A, Trewhela A, Watt I. Yoga for chronic low back pain a randomized trial. Ann Intern Med. 2011;155(9):569–78.
24. Williams K, Abildso C, Steinberg L, Doyle E, Epstein B, Smith D, Hobbs G, Gross R, Kelley G, Cooper L. Evaluation of the effectiveness and efficacy of Iyengar yoga therapy on chronic low back pain. Spine. 2009;34(19):2066–76.
25. Groessl EJ, Weingart KR, Johnson N, Baxi S. The benefits of yoga for women veterans with chronic low back pain. J Altern Complement Med. 2012;18(9):832–8.
26. Chuang L-H, Soares MO, Tilbrook H, Cox H, Hewitt CE, Aplin J, Semlyen A, Trewhela A, Watt I, Torgerson DJ. A pragmatic multicentered randomized controlled trial of yoga for chronic low back pain: economic evaluation. Spine. 2012;37(18):1593–601.
27. Michalsen A, Traitteur H, Lüdtke R, Brunnhuber S, Meier L, Jeitler M, Büssing A, Kessler C. Yoga for chronic neck pain: a pilot randomized controlled clinical trial. J Pain. 2012;13(11):1122–30.

28. Telles S, Dash M, Naveen K. Effect of yoga on musculoskeletal discomfort and motor functions in professional computer users. Work. 2009;33(3):297–306.
29. Omkar S, Mour M, Das D. A mathematical model of effects on specific joints during practice of the Sun Salutation–a sequence of yoga postures. J Bodyw Mov Ther. 2011;15(2):201–8.
30. Hart CE, Tracy BL. Yoga as steadiness training: effects on motor variability in young adults. J Strength Cond Res. 2008;22(5):1659–69.
31. Ülger Ö, Yağlı NV. Effects of yoga on balance and gait properties in women with musculoskeletal problems: a pilot study. Complement Ther Clin Pract. 2011;17(1):13–5.
32. Zettergren KK, Lubeski JM, Viverito JM. Effects of a yoga program on postural control, mobility, and gait speed in community-living older adults: a pilot study. J Geriatr Phys Ther. 2011; 34(2):88.
33. Gonçalves LC, Vale RG, Barata NJF, Varejão RV, Dantas EHM. Flexibility, functional autonomy and quality of life (QoL) in elderly yoga practitioners. Arch Gerontol Geriatr. 2011;53(2): 158–62.
34. Chen K-M, Fan J-T, Wang H-H, Wu S-J, Li C-H, Lin H-S. Silver yoga exercises improved physical fitness of transitional frail elders. Nurs Res. 2010;59(5):364–70.
35. Schmid AA, Van Puymbroeck M, Koceja DM. Effect of a 12-week yoga intervention on fear of falling and balance in older adults: a pilot study. Arch Phys Med Rehabil. 2010;91(4): 576–83.
36. DiBenedetto M, Innes KE, Taylor AG, Rodeheaver PF, Boxer JA, Wright HJ, Kerrigan DC. Effect of a gentle Iyengar yoga program on gait in the elderly: an exploratory study. Arch Phys Med Rehabil. 2005;86(9):1830–7.
37. Schmid AA, Van Puymbroeck M, Altenburger PA, Schalk NL, Dierks TA, Miller KK, Damush TM, Bravata DM, Williams LS. Poststroke balance improves with yoga a pilot study. Stroke. 2012;43(9):2402–7.
38. Greendale GA, Huang MH, Karlamangla AS, Seeger L, Crawford S. Yoga decreases kyphosis in senior women and men with adult-onset hyperkyphosis: results of a randomized controlled trial. J Am Geriatr Soc. 2009;57(9):1569–79.

Index

E. Mody, E. Matzkin (eds.), *Musculoskeletal Health in Women*,
DOI 10.1007/978-1-4471-4712-1, © Springer-Verlag London 2014

MIX
Papier aus verantwortungsvollen Quellen
Paper from responsible sources
FSC® C105338

If you have any concerns about our products,
you can contact us on
ProductSafety@springernature.com

In case Publisher is established outside the EU,
the EU authorized representative is:
Springer Nature Customer Service Center GmbH
Europaplatz 3, 69115 Heidelberg, Germany

Printed by Libri Plureos GmbH
in Hamburg, Germany